D1140053

Performance Interventions

Series Editors: Elaine Aston, University of Lancaster, and Bryan Reynolds, University of California, Irvine

Performance Interventions is a series of monographs and essay collections on theatre, performance, and visual culture that share an underlying commitment to the radical and political potential of the arts in our contemporary moment, or give consideration to performance and to visual culture from the past deemed crucial to a social and political present. *Performance Interventions* moves transversally across artistic and ideological boundaries to publish work that promotes dialogue between practitioners and academics, and interactions between performance communities, educational institutions, and academic disciplines.

Titles include:

Alan Ackerman and Martin Puchner (*editors*)
AGAINST THEATRE
Creative Destructions on the Modernist Stage

Elaine Aston and Geraldine Harris (*editors*)
FEMINIST FUTURES?
Theatre, Performance, Theory

Maaike Bleeker
VISUALITY IN THE THEATRE
The Locus of Looking

James Frieze
NAMING THEATRE
Demonstrative Diagnosis in Performance

Lynette Goddard
STAGING BLACK FEMINISMS
Identity, Politics, Performance

Alison Forsyth and Chris Megson (*editors*)
GET REAL: DOCUMENTARY THEATRE PAST AND PRESENT

Leslie Hill and Helen Paris (*editors*)
PERFORMANCE AND PLACE

D.J. Hopkins, Shelley Orr and Kim Solga
PERFORMANCE AND THE CITY

Amelia Howe Kritzer
POLITICAL THEATRE IN POST-THATCHER BRITAIN
New Writing: 1995–2005

Jon McKenzie, Heike Roms and C.J.W.-L. Wee (*editors*)
CONTESTING PERFORMANCE
Global Sites of Research

Melissa Sihra (*editor*)
WOMEN IN IRISH DRAMA
A Century of Authorship and Representation

LIVERPOOL JMU LIBRARY

3 1111 01418 0333

Performance Interventions
Series Standing Order ISBN 978–1–4039–4443–6 Hardback
978–1–4039–4444–3 Paperback
(outside North America only)

You can receive future titles in this series as they are published by placing a standing order. Please contact your bookseller or, in case of difficulty, write to us at the address below with your name and address, the title of the series and the ISBN quoted above.

Customer Services Department, Macmillan Distribution Ltd, Houndmills, Basingstoke, Hampshire RG21 6XS, England

Contesting Performance

Global Sites of Research

Edited by

Jon McKenzie
Associate Professor of English, University of Wisconsin, USA

Heike Roms
Senior Lecturer in Performance Studies, Aberystwyth University, UK

C.J.W.-L. Wee
Associate Professor of English, Nanyang Technological University, Singapore

Introduction, selection and editorial matter © Jon McKenzie,
Heike Roms & C. J. W.-L. Wee 2009, 2012
Individual chapters © contributors 2009, 2012
Foreword © Maaike Bleeker 2012

All rights reserved. No reproduction, copy or transmission of this
publication may be made without written permission.

No portion of this publication may be reproduced, copied or transmitted
save with written permission or in accordance with the provisions of the
Copyright, Designs and Patents Act 1988, or under the terms of any licence
permitting limited copying issued by the Copyright Licensing Agency,
Saffron House, 6–10 Kirby Street, London EC1N 8TS.

Any person who does any unauthorized act in relation to this publication
may be liable to criminal prosecution and civil claims for damages.

The authors have asserted their rights to be identified
as the authors of this work in accordance with the Copyright, Designs
and Patents Act 1988.

First published in hardback 2009 in
Published in paperback 2012 by
PALGRAVE MACMILLAN

Palgrave Macmillan in the UK is an imprint of Macmillan Publishers Limited,
registered in England, company number 785998, of Houndmills, Basingstoke,
Hampshire RG21 6XS.

Palgrave Macmillan in the US is a division of St Martin's Press LLC,
175 Fifth Avenue, New York, NY 10010.

Palgrave Macmillan is the global academic imprint of the above companies
and has companies and representatives throughout the world.

Palgrave® and Macmillan® are registered trademarks in the United States,
the United Kingdom, Europe and other countries.

ISBN 978–0–230–00845–8   hardback
ISBN 978–1–137–01118–3   paperback

This book is printed on paper suitable for recycling and made from fully
managed and sustained forest sources. Logging, pulping and manufacturing
processes are expected to conform to the environmental regulations of the
country of origin.

A catalogue record for this book is available from the British Library.

A catalog record for this book is available from the Library of Congress.

10   9   8   7   6   5   4   3   2   1
21   20   19   18   17   16   15   14   13   12

Printed and bound in Great Britain by
CPI Antony Rowe, Chippenham and Eastbourne

Contents

Foreword to the Paperback Edition

This book provides a much needed introduction to the diversity of performance research as it takes place in various places in the world. This introduction is, of course, as the editors rightfully observe, only partial. There are too many developments and situations to be included in a single volume. More than that, one might argue that this partiality is precisely a condition of the situation to which this volume presents an introduction. It is only by acknowledging the impossibility of a perspective that would be all inclusive that we can begin to think of the complexities of performance research in a global situation and fully appreciate the potential of this situation.

This volume addresses these complexities first of all by means of a focus on the 'situatedness' of local sites of research. The essays brought together in this book provide wonderful insights into the specificities of various sites of research and identify, as the editors put it, 'alternative genealogies to a research paradigm long associated with the New York University's (NYU) Department of Performance Studies' (p. 4). The NYU paradigm is a recurring motif in many contributions to this volume, and as a historical point of reference this centrality is not surprising, given the fact that this was the first one to be formally institutionalized. The NYU research paradigm has come to represent the US approach to performance studies (even though, as Shannon Jackson in this volume demonstrates, other paradigms exist also within the US) and has been exported, sometimes more, sometimes less successfully, to other parts of the world. The essays collected in this volume demonstrate how this approach has been an important source of both inspiration and resistance, and how this resistance teams up with the resistance to other US 'export products' like neoliberal capitalism, the US model of democracy, and the so called global war on terror (to mention just a few), all of them presupposing a universally shared perspective that justifies what is in fact the convictions and interests of a particular 'situatedness' (and even within this situation represents only a very partial view that is by far not shared by everyone). Here I would like to remind us of Susan Buck-Morss' observations (in her *Thinking Past Terror*) that: 'Global space entails simultaneity, overlap, coherencies incoherently superimposed. Like a photograph in multiple exposure, it makes sense only precariously, only by blocking out part of the visible field. We are capable of seeing further than is comprehended by our separate, sense-making practices, and what we see limits the legitimacy of what we do' (Buck-Morss 2003, 5).

The essays collected in this volume demonstrate that alternative paradigms exist that can be traced back to different traditions and practices in different local sites of research. And that the becoming global of performance research is not in the first place a matter of the cultural phenomena that are the subject of study, nor of the world-wide incorporation of one particular research paradigm, but rather of the diversification of local perspectives implied within modes of approaching and understanding the phenomena studied. Today, as the editors point out, 'performance studies is no longer only about the West—specifically the United States—studying the "Rest"' (p. 2) but about contesting approaches to performance: 'While performance has for some time been recognized as both a contested concept and a practice of potential contestation, the sites and stakes of those contests have both multiplied and entered into new configurations' (p. 2). A crucial question therefore is how to engage with this multiplication of perspectives and multiple sites of contestation, and how to make this situation of 'multiple exposure' (Buck-Morss) productive for the further development of the field.

One month prior to the time of writing this Foreword, the future development of the field of performance research was the subject of a conference organized at Princeton. Titled *Performance Studies: Memories and Futures*, this conference paid a well-deserved tribute to Richard Schechner as the founding father of the NYU paradigm. Rather disturbing, however, especially for many performance scholars outside the US, was that the list of speakers was exclusively made up of performance scholars affiliated with US universities (with the exception of one Canadian scholar). Not only the memories of Performance Studies past, but also the future of Performance Studies (PS) were thus presented as an exclusively US enterprise. None of the editors or contributors to this volume were included in the list of speakers, nor were other prominent PS scholars from the international field, not even the US/UK/Australian former presidents of Performance Studies international (PSi), the organization representing the field of Performance Studies internationally.

All of this happened at the same time that within PSi a shift was taking place away from US dominance towards an increased inclusion of the rest of the world. While the size of the annual PSi conference is expanding and the number of scholars and artists involved with PSi is growing, the board of the organization which represents this field also has become more diverse in terms of background and affiliation. At the time of writing, only four of the 20 board members of PSi are affiliated with a US university, and PSi has, in myself, its first non-US/UK/Australian president and its first president who is not a native speaker of English. The ambition to become more internationally diverse, as well as an awareness of the difficulties related to this ambition, are part of the very foundations of the organization and reflected in its name. The small 'i' stands for the ambition to be international in a way that is decentralized, and for Performance Studies that are not one but many. Such being international is a project that is difficult because it requires willingness to accept an identity that is built on constantly shifting ground, and because of the lack of a shared, unifying frame of reference. This situation requires from us willingness to accept that to perform such internationalism is not merely a matter of opening up to others to include them in our own paradigm, but also involves making the position of the centre as – to speak with Peggy Phelan – unmarked. Being international involves a reconsideration of Performance Studies as a conglomerate of culturally specific, or, as the editors of this volume put it, situated phenomena.

What does it mean that the international field of performance research – and the PSi organization that represents this field – is excluded from the memories and future of Performance Studies as staged at Princeton? Of course there may have been pragmatic reasons, like, for example, a lack of financial means, to explain the absence of international guests. However, what matters is a certain self-reflexivity, or the absence of such, with regard to the claim performed by the framing of what was offered, and, in particular, how this framing leaves the situatedness of what is presented unmarked. It would have been an entirely different situation if this line-up of speakers had been framed as 'Performance Studies in the US: Memories and Futures'. This way, the very same line-up could have highlighted the specificity of the perspective presented, and how this perspective is related to the US situation. It could also have contributed to a further renegotiation of the relationships between performance research as a global phenomenon and the specificities of a US approach to Performance Studies.

Useful here is the notion of the 'former West' as proposed by Maria Hlavajova of the Utrecht Visual Arts Centre BAK. Hlavajova proposes 'former West' as a counter-narrative to the 'former East' as both an ideological and geographical space constructed in the rhetoric of Western democracies after the fall of the Berlin Wall. That a concomitant shift from the 'West' to the 'former West' was never articulated, Hlavajova argues, points to the fact that the societies of Western Europe and the

US failed to confront what implications the collapse of communism in Europe had for themselves.

> The so-called West, blinded by the (default) victory of neoliberal cap-italism on a global scale, failed to recognize the impact of the massive shifts put into motion by the events of that year, and has continued to adhere to its own claims of hegemony. The term 'former West,' never articulated as a counterpart to the widely used 'former East,' thus does not refer to the status quo, but is rather an aspired to, imag-ined 'farewell' to the 'bloc' mentality; it is a critical, emancipatory, and aspirational proposal to rethink our global histories and to spec-ulate upon our global futures through artistic and cultural practice. (http://www.formerwest.org/About)

What could it mean to consider Performance Studies a former West-ern invention? Following Hlavajova, a first step would be to rethink what used to be the West as itself a multiplicity of regions among other regions. This involves acknowledging the situatedness not only of alter-native knowledges and paradigms, but also of those associated with what used to be the centre. This volume presents a great example of doing so. Not only is the reader introduced to many 'alternative genealogies' of per-formance research, as the editors put it, but these are presented side by side with chapters discussing the situatedness of performance research in the US, UK, and Australia, i.e., countries that are usually associated with the centre rather than with the periphery. The book thus presents them as regions among other regions, while the individual chapters about these regions draw attention to the cultural specificity of performance research as it developed in these regions. Furthermore, these chapters show the development of performance research in these regions (including those considered to be part of or close to the centre) to be marked by contesting approaches, thus further destabilizing the oppositions of self and other, centre and periphery. Australia is represented by two entirely different chapters, one by Edward Scheer and Peter Eckersall, and another one by Gay McAuley. And Shannon Jackson's chapter on the US does not engage with the NYU paradigm but rather with its 'other', the oral, rhetorical tradition of performance research, and focuses on the part performed by this other tradition in the institutionalization of performance studies in the US.

Highlighting the specificities of performance research in these 'regions among regions', the contributions to this volume draw attention to both the potential of specific practices of research to make things visible and to how such making visible also results in the 'blocking out' of others. Several authors reflect on the ways in which the opposition of theatre and performance, instrumental to the institutionalization of US performance studies, works out entirely differently in other contexts. Not only are performative practices as well as research into performance organized along other lines of distinction, the development of ideas related to this distinction between theatre and performance happens according to a chronology that undermines the idea of a global unified history. For example, Lada Čale Feldman and Marin Blažević in their contribution point out how ideas that within the US performance studies paradigm are associated with the way performance liberates itself from the constraints of theatrical representation, in Croatia actually appear much earlier (prior to the emergence of performance and performance studies) and within the discourse surrounding theatre. Their contribution (as well as other contributions to this volume) thus invites, following Hlavajova, a rethinking of global history beyond the chronology suggested by the former Western unitary perspective.

I am writing these reflections from a New York apartment located at a walking distance from New York University's Department of Performance Studies. I have just arrived here for a six-month period as visiting scholar at NYU. I received my training far from NYU, both geographically and ideologically. Although the country where I was born and grew up is certainly not on the margin of the neoliberal economy and is firmly part of the former West, when it comes to the NYU performance studies paradigm, the Netherlands belongs to the margin. My training was in Theatre Studies, Art History, and Philosophy. I obtained my PhD from the Amsterdam School for Cultural Analysis. The word 'performance' as used in English does not translate into Dutch, and in the Dutch context the opposition of theatre and performance does not make sense the way it does in an Anglo-American context. In many ways, my situation is representative (to a greater or lesser degree) of quite a few of the alternative genealogies discussed on the pages that follow. And it is from this position that I am now representing Performance Studies internationally as PSi's president. It is also from this position that I am now bringing my alternative genealogy to NYU, eager to do what this book invites us to do, namely to bring the specificity of our situated practices into a situation of exchange and mutual inspiration from where we may speculate upon a global future.

Maaike Bleeker

Acknowledgements

Putting together this collection has taken far more time and effort than we ever imagined it would. Working as editors who live on or near three distant continents, organizing our efforts across disjointed time zones, meeting face-to-face only at far-flung conferences, diners, and coffee shops – all of this has given us grudging respect for the collaborative efficiencies of multinational corporations. It has all come together – finally.

For this coming together, we wish to thank first of all our authors for their work in meeting all (or most) of the deadlines we posed, for their patience when these deadlines were extended, for their understanding when there were misunderstandings, and, most of all, for their fine contributions to this book.

We also thank participants and organizers of the academic events that helped shape our work and our thinking, specifically those of the 'Internationalism and Performance Studies' symposium held at the 2003 New York meeting of the Association of Theatre in Higher Education; the 'Glocalizing Performance' workshop held at the 2004 Singapore conference of Performance Studies international; and 'The Stakes of Performance Research' seminar held at the Chicago 2006 conference of the American Society for Theatre Research. Special thanks goes to Jessica Chalmers for co-organizing the 'Internationalism and Performance Studies' symposium.

Along the way, numerous scholars have offered advice, suggestions, and criticisms of this project. We wish to thank in particular Rustom Bharucha, Josette Féral, Ric Knowles, Goenawan Mohamad, Richard Schechner, and Diana Taylor. Special thanks to Caroline Levine for her suggestions and input, and to Richard Gough, Stephen Bottoms and *Performance Research* for supporting the publication of Shannon Jackson's essay here. And our very special thanks to Maaike Bleeker for contributing a new Foreword to this edition.

Our project received much intellectual encouragement from the Performance Intervention series editors, Elaine Aston and Bryan Reynolds. Martin Puchner likewise provided early editorial support and guidance. Our senior editor at Palgrave Macmillan, Paula Kennedy, ably kept us going even when things had almost ground to a halt. Penny Simmons, our editorial consultant, provided timely work on the text, and our indexer Joshua Taft came through when things counted most – at the end.

Notes on the Contributors

Khalid Amine is Senior Professor of Comparative Literature and Performance Studies at the English Department, Faculty of Humanities, Tétouan, Morocco. He is the Founding President of The International Centre for Performance Studies (Tangier), Research Fellow at the Institute for Interweaving Performance Cultures (Free University Berlin, 2008–10), and winner of the 2007 Helsinki Prize of the International Federation for Theatre Research (FIRT). Among his publications is *Dramatic Art and the Myth of Origins: Fields of Silence* (International Centre for Performance Studies Publications, 2007), and his essays have appeared in *The Drama Review (TDR)*, *Critical Survey*, and *Theatre Journal*.

Sharon Aronson-Lehavi is Assistant Professor of Theatre Studies at the Department of Comparative Literature, Bar Ilan University, Israel. She holds a PhD in Theatre Studies from the Graduate Center, City University of New York, USA. She is a Fulbright grantee and a winner of a Dan David award for postdoctoral studies. She is the author of *Street Scenes: Late Medieval Acting and Performance* (Palgrave Macmillan, 2011) and the editor of *Wanderers and Other Israeli Plays* (Seagull Books, 2009). Her essays have appeared in *TRI*, *Performance Research* and other journals and essay collections.

Marin Blažević is Assistant Professor in the Department of Dramaturgy at the Academy of Drama Arts, University of Zagreb, Croatia. He was editor-in-chief of the performing arts journal *Frakcija*, editor of the *Akcija (Action)* book series on performing arts and performance theory, and director of the 15th Performance Studies international conference (Zagreb, 2009). He co-edited *Branko Gavella: Teorija glume – od materijala do ličnosti* (Branko Gavella: Theory of Acting – From Material to Personality) (CDU, 2005), and, with Matthew Goulish, *Reflections on the Process/Performance: A Reading Companion to Goat Island's 'When will the September roses bloom? Last night was only a comedy'* (*Frakcija*, 2004–05). Together with Lada Čale Feldman he is currently working on a collection of essays, *MISperforming: Inverting, Shifting, Failing*, to be published in English in 2012. His monographs include *Razgovori o novom kazalištu* (*Conversations on the New Theater*) published in 2007, and *Izboren poraz* (*A Defeat Won*), on the theory of new theatre and its peculiar history in Croatia, to be published in Spring 2012.

Maaike Bleeker is Professor of Theatre Studies at Utrecht University, The Netherlands. Since 1991, she has combined her academic work with a practice as dramaturg, collaborating with various theatre directors, choreographers, and visual artists. She was the organizer of the 2011 world conference of Performance Studies international, titled Camillo 2.0: Technology, Memory, Experience (www.psi17.org). She is the author of *Visuality in the Theatre: The Locus of Looking* (Palgrave Macmillan, 2008) and *Anatomy Live: Performance and the Operating Theatre* (Amsterdam University Press, 2008). She has been President of Performance Studies international (PSi) since May 2011.

Peter Eckersall is Associate Professor of Theatre Studies in the School of Culture and Communication, University of Melbourne, Australia. His research interests include contemporary Japanese theatre and culture, experimental performance and dramaturgy. Recent publications include *Theorising the Angura Space: Avant-garde Performance and Politics in Japan 1960–2000* (Brill Academic, 2006) and *Kawamura Takeshi's Nippon Wars and Other Plays* (Seagull Books, 2010). He is currently working on *Revolution and the Everyday: Performative Interactions in Art, Theatre and Politics in 1960's Japan* and a comparative study of modernism and theatre in the Asia-Pacific region. Peter was chair of the international committee of PSi (2006–10) and is co-founder and co-editor of *Performance Paradigm*.

Lada Čale Feldman is Professor in the Department for Comparative Literature at the University of Zagreb, where she teaches drama, theatre and performance studies. As a former research associate in the Institute of Ethnology and Folklore Research, her areas of interest also include folk theatre, political propaganda, and gender studies. Her publications include *Bresanov teatar* (*Bresan's Theatre*) (Hrvatsko društvo kazališnih kritičara i teatrologa, 1989), *Teatar u teatru u hrvatskom teatru* (*Play-within-the-Play in the Croatian Theatre*) (Naklada MD, 1997), *Euridikini osvrti* (Eurydice's Turns) (Centar za ženske studije i Naklada MD, 2001), for which she won the Petar Brecic Award, *Femina ludens* (Disput, 2005) and, co-authored with M. Čale, *U kanonu* (*In the Canon*) (Disput, 2008). She also co-edited (with I. Prica and R. Senjković) *Fear, Death and Resistance – An ethnography of War, Croatia 1991–1992* (Institute of Ethnology and Folklore Research/Matrix Croatic, 1993) and (with I. Prica) *Etnografija domaceg socijalizma* (Institut za etnologiju i folkloristiku, 2006).

Shannon Jackson is Richard and Rhoda Goldman Distinguished Professor in the Arts and Humanities and Director of the Arts Research Center at the University of California, Berkeley, USA. She teaches regularly in performance theory, theatre, and the visual arts. Her publications include *Lines of Activity: Performance, Historiography, and Hull-House Domesticity* (University of Michigan Press, 2000), *Professing Performance: Theatre in the Academy from Philology to Performativity* (Cambridge University Press, 2004), *Social Works: Performing Art, Supporting Publics* (Routledge, 2011), and numerous essays, published in journals including *The Drama Review (TDR)*, *The Journal of Visual Culture*, *Theatre Survey*, *Cultural Studies*, *Modern Drama*, *Theatre Journal*, *Theatre Topics*, *Text and Performance Quarterly*, and numerous collections. She is presently completing, with Marianne Weems, *The Builders Association* (MIT Press, forthcoming).

Loren Kruger is Professor of Comparative and English Literatures, African Studies, and Theatre and Performance Studies at the University of Chicago, USA. She is the author of *The National Stage* (University of Chicago Press, 1992), *The Drama of South Africa* (Routledge, 1999), and *Post-Imperial Brecht* (Cambridge University Press, 2005), for which she won the Scaglione Prize for Comparative Literature awarded by the Modern Language Association. Her articles have appeared in many international journals including *Diaspora*, *Frakcija*, *Journal of Southern African Studies*, *Poetics Today*, *Theater*, *Theater der Zeit* and *The Drama Review (TDR)*. She is a contributing editor of *Theatre Research International*, and advisory board member for *Modern Drama* and *Scrutiny 2*.

Bojana Kunst is a philosopher, dramaturg, teacher, and performance theoretician. She works as guest professor at the University of Hamburg (Performance Studies). She is a member of the editorial boards of *Maska*, *Amfiteater* and *Performance Research*, and her essays have appeared in numerous journals and collections. Her books include *Impossible Body* (Maska, 1999), *Dangerous Connections: Body, Philosophy and Relation to the Artificial* (Maska, 2004) and *Processes of Work and Collaboration in Contemporary Performance* (Maska/Amfiteater, 2010).

Ray Langenbach (raylangenbach@mac.com). Langenbach's visual art and performances have been presented in the United States, Europe, and the Asia-Pacific. His writings on Southeast Asian performance, propaganda, and visual culture have appeared in *Performance Research*, *Afterimage*, *Oxford Dictionary of Performance*, and *Eye of the Beholder: Reception, Audience and Practice of Modern Asian Art*, ed. John Clark, Maurizio Peleggi and T. K. Sabapathy (Wild Peony, 2006), among others. He co-convened Performance Studies international's tenth conference (Singapore, 2004), Satu Kali International Performance Art Symposium (Kuala Lumpur, 2006) and curated the Performance Art section at the

2000 Werkleitz Biennial. Langenbach is Star Foundation Research Chair Professor in the Faculty of Creative Industries, Universiti Tunku Abdul Rahman, Malaysia, and is also a Research Fellow at the Finnish Academy of Fine Art in Helsinki, Finland.

Gay McAuley is an Honorary Associate Professor in the Department of Performance Studies at the University of Sydney, Australia. She taught theatre and film in the French Department before establishing Performance Studies as an interdisciplinary centre in 1989. Her book *Space in Performance* (University of Michigan Press, 1999) examines the many functions of space in the theatre experience; she then extended her exploration of spatial semiotics to site-based performance practices, with particular reference to the relation between place and memory (*Unstable Ground: Performance and the Politics of Place*, Peter Lang, 2006). Her current research concerns creative agency in the rehearsal process.

Jon McKenzie is Director of DesignLab and Associate Professor of English at the University of Wisconsin-Madison, where he teaches courses in performance theory, new media, and civil disobedience. In addition, he coordinates an interdisciplinary initiative in digital humanities involving new media studies, studio-based practices, digital learning, and quantitative humanities research. McKenzie is author of *Perform or Else: From Discipline to Performance* (Routledge, 2001), and such articles as 'Democracy's Performance', 'Global Feeling: (Almost) All You Need is Love', 'High Performance Schooling', 'StudioLab UMBRELLA', and 'Abu Ghraib and the Society of the Spectacle of the Scaffold'. His work has appeared in such journals as *The Drama Review (TDR)*, *Performance Research*, and *Parallax*, and has been translated into numerous languages, including Croatian, French, German, Polish, Portuguese, and Spanish. He co-edited a special issue of the performance journal *Frakcija* on the topic of security, visibility, and civil liberty.

Sal Murgiyanto is Associate Professor of Dance and Performance at the Taipei National University of the Arts, Taiwan, and the Jakarta Institute of the Arts, Indonesia. He earned his BA (1975) from ASTI National Dance Academy of Indonesia in Yogyakarta; his MA (Dance, 1976) from the University of Colorado; and his PhD (Performance Studies, 1991) from New York University, USA. He is the founder and on the artistic board of the Indonesian Dance Festival (1992–), and also the founder and on the board of the MSPI Society for the Indonesian Performing Arts. Recent publications include: 'Sardono: Dialogues with Humankind and Nature,' in *Dance, Human Right, and Social Justice,* ed. Naomi Jackson and Toni Shapiro-Phim (Scarecrow Press, 2008); 'From Village to Theatrical Stage and Back', *Asia-Pacific Forum* 39 (2008); and 'Reinventing Tradition: New Dance in Indonesia', in *Shifting Sands: Dance in Asia and the Pacific,* ed. Stephanie Burridge (Ausdance National, 2006).

Sibylle Peters is a researcher, director, and performer. She studied cultural studies and philosophy, and worked as researcher and lecturer at the Universities of Hamburg, Munich, Berlin (Germany), and Bale (Switzerland). As a director she has realized performance projects that are concerned primarily with questions of participation and collective research, often in cooperation with performance collective geheimagentur. Peters founded the Forschungstheater im FUNDUS THEATER in Hamburg, a place where children, artists, and scientists meet. She is a member of the board of directors of FUNDUS THEATER and of the graduate college Versammlung und Teilhabe at the HafenCity University Hamburg. Recent publications include *Szenen des Vorhangs – Schnittflächen der Künste,* ed. with G. Brandstetter (Rombach, 2008), and *Der Vortrag als Performance* (Transcript, 2011) on the academic lecture as performance.

Paul Rae is Assistant Professor in the Theatre Studies Programme at the National University of Singapore. He is the author of *Theatre and Human Rights* (Palgrave Macmillan, 2009), and his work on cosmopolitanism, mobility and contemporary Southeast Asian performance has appeared in journals such as *The Drama Review (TDR), Contemporary Theatre Review,* and *Performance Research.* He is the co-artistic director of spell#7 (www.spell7.net), and, in 2004, worked with other Singapore-based artists and critics to host the tenth annual conference of Performance Studies international (PSi).

Freddie Rokem is the Emanuel Herzikowitz Professor for 19th and 20th Century Art and teaches in the Department of Theatre Studies at Tel Aviv University (Israel), where he served as the Dean of the Yolanda and David Katz Faculty of the Arts (2002–06). He is also a permanent Visiting Professor at Helsinki University, Finland. He has been a visiting Professor at the Universities of Munich and Stockholm, Stanford University, the Free University in Berlin, and the University of California, Berkeley. He was editor of *Theatre Research International* (2006–09). Rokem's book *Performing History: Theatrical Representations of the Past in Contemporary Theatre* (University of Iowa Press, 2000) received the ATHE (Association for Theatre in Higher Education) Prize for best theatre studies book in 2001. *Strindberg's Secret Codes* was published by Norvik Press (2004) and *Philosophers and Thespians: Thinking Performance* by Stanford University Press (2010).

Heike Roms is Senior Lecturer in Performance Studies at Aberystwyth University, UK. Originally from Germany, she moved to Wales in 1995 and became the first administrator of PSi (Performance Studies international) (1998–2001) and co-organiser of the 5th Performance Studies Conference at Aberystwyth in 1999. She has published widely on contemporary performance, in particular on work emanating from Wales, for publications such as *Performance Research, Frakcija, Inter, Cyfrwng, Ballett/Tanz*, and a number of collections. She is a contributing editor of *Performance Research* and serves on the editorial boards of *Frakcija* and *Inter.* Heike's current research project, 'Locating the early history of performance art in Wales 1965–1979' (www.performance-wales.org), focuses on the historiography of performance art. The project was funded by a Large Research Grant from the British Arts and Humanities Research Council AHRC (2009–11) and won the David Bradby TaPRA Award for Outstanding Research in International Theatre and Performance 2011.

Edward Scheer is Professor in the School of English, Media and Performing Arts at the University of New South Wales, Sydney, Australia. He is a founding editor of the journal *Performance Paradigm* with Peter Eckersall. His study of duration in Mike Parr's performance art, *The Infinity Machine*, was published by Schwartz Press in 2009. He has edited two books on Artaud – *100 Years of Cruelty: Essays on Artaud* (Artspace and Power Publications, 2000) and *Antonin Artaud: A Critical Reader* (Routledge, 2004). He is co-editor with Peter Eckersall of *The Ends of the 60s: Performance, Media and Contemporary Culture* (Performance Paradigm, 2006) and with John Potts of *Technologies of Magic: A Cultural Study of Ghosts, Machines and the Uncanny* (Power Publications, 2006). He served as chairman of the board of directors of the Performance Space in Sydney from 2005–07. Scheer was President of PSi (Performance Studies international) (2007–11).

Takahashi Yuichiro is Professor of Performance Studies in the Department of Tourism and Transnational Studies, Dokkyo University, Japan. Takahashi is the author of *Shintai-ka Suru Chi* (*Embodied Knowledge*) (Serica Shobo, 2005). His articles have been anthologized in *Alternatives: Debating Theatre Culture in the Age of Con-Fusion*, ed. Peter Eckersall, Uchino Tadashi and Moriyama Naoto (P.I.E.-Peter Lang, 2004) and *A Kabuki Reader*, ed. Samuel L. Leiter (M. E. Sharpe, 2002) and have appeared in *The Drama Review (TDR)*. His translation (into Japanese) includes books, articles, and poems by Richard Schechner, Paul Bowles, Jack Kerouac, and Alice B. Toklas.

Diana Taylor is University Professor and Professor of Performance Studies and Spanish at New York University (USA) and Director of the Hemispheric Institute of Performance and Politics. She is the author of *Theatre of Crisis: Drama and Politics in Latin America* (University Press of Kentucky, 1991), *Disappearing Acts: Spectacles of Gender and Nationalism in Argentina's 'Dirty War'* (Duke University Press, 1997) and *The Archive and the Repertoire: Performing Cultural Memory in the Americas* (Duke University Press, 2003), which won the Outstanding Book from ATHE (Association for Theatre in Higher Education) and the Katherine Singer Kovacs Prize from the Modern Language Association.

Uchino Tadashi is Professor of Performance Studies in the Department of Interdisciplinary Cultural Studies, Graduate School of Arts and Sciences, University of Tokyo, Japan. He received his MA in American Literature (1984) and his PhD in Performance Studies (2001) from the University of Tokyo. His research interest includes contemporary Japanese and American theatre and performance, and his publications include: *The Melodramatic Revenge: Theatre of the Private in the 1980s* (in Japanese; Keiso-shobo, 1996), *From Melodrama to Performance: The Twentieth Century American Theatre* (in Japanese; UT Press, 2001), *Crucible Bodies: Postwar Japanese Performance from Brecht to the New Millennium* (Seagull Press, 2009) and *Perspectives from the Stage: Tokyo/New York 1995–2005*, two volumes (in Japanese; Renga-shobo Shin-sha, 2010). He is a contributing editor for *The Drama Review (TDR)*, an editor for *Performing Arts* (Kyoto University of Arts and Design), and the *Journal of American Literature Studies in Japan*.

C. J. W.-L. Wee is Associate Professor of English at the Nanyang Technological University, Singapore. He has held Visiting Fellowships at the Humanities Research Centre, Australian National University, and at the Society for the Humanities, Cornell University (USA). Wee is the author of *Culture, Empire, and the Question of Being Modern* (Lexington, 2003) and *The Asian Modern: Culture, Capitalist Development, Singapore* (Hong Kong University Press, 2007), he is also the editor of *Local Cultures and the 'New Asia': The State, Culture, and Capitalism in Southeast Asia* (Institute of Southeast Asian Studies, 2002). His essays have appeared in journals such as *Public Culture, Critical Inquiry, The Drama Review (TDR)*, and *positions: east asia cultures critique*. His present research interest is in the contemporary arts, literature, and the culture industries in East Asia, and the relationship between questions of the postcolonial, modernity/modernism, and the contemporary.

Introduction: Contesting Performance in an Age of Globalization

Jon McKenzie, C. J. W.-L. Wee, and Heike Roms

Performance research has gone global. By this we refer not so much to the cultural phenomena studied – which, it is clear, have long been located around the world – but to the *locations* of researchers themselves. These locations have steadily expanded over the past two decades, whether it be in terms of individual researchers working alone or in small groups on different continents. This expansion is mirrored by the emergence of performance research and study programs in different countries. While the United States continues to host many influential scholars and programs, the United Kingdom in particular has seen an increase in performance scholarship and in university courses of study that carry the term 'performance' in their names, and important research projects and academic departments have emerged in locales as diverse as Australia, Brazil, Canada, China, Croatia, Denmark, France, Germany, India, Indonesia, Israel, Japan, Mexico, Morocco, the Netherlands, New Zealand, Nigeria, Peru, Poland, Singapore, Slovenia, and South Africa. In addition, a number of transnational scholarly organizations have formed – some with a regional focus, notably the Hemispheric Institute of Performance and Politics and the Asian Performance Studies Research Group, and others with an international scope, such as Performance Studies international (PSi) and the performance-focused working groups of the International Federation of Theatre Research (IFTR). The Centre for Performance Research was established in the United Kingdom in 1988, and in 2005, the Schechner Center for Performance Studies opened at the Shanghai Theatre Academy in China.

If performance research as a recognized area of study and its institutionalization as performance studies have been widely perceived as centered in the United States, there is, also, a growing sense that a profound decentering of the area is transpiring, one that this collection seeks

1

to register and document. Performance studies is no longer only about the West – specifically the United States – studying the 'Rest.' While performance has for some time been recognized as both a contested concept and a practice of potential contestation, the sites and stakes of those contests have both multiplied and entered into new configurations.

This collection, *Contesting Performance: Global Sites of Research*, has three major aims. The first is to foreground diverse locations of research, offering a partial survey of the globalization of performance research – and we use the terms 'partial' and 'globalization' advisedly here. The volume is partial for, in the end, there are too many developments and situations to be included in a single volume. Several important geographical areas are not represented, for instance, Canada, China, France, and India, though performance research is produced there. The chapters collected here are also partial in that each embodies particular local perspectives and – even when local complexities justify joint authorship for a single chapter – does not attempt universalizing summations, even of its own situation.

Indeed, while performance research has gone global, this development has not been altogether positive. As several contributions to the volume contend, such research has sometimes coincided in troubling ways with neoliberal economic globalization. Despite the critical attitude toward economic globalization taken by many, if not most performance researchers, the globalizing processes of performance and neoliberal economics are not unrelated or opposed, given the burgeoning global market for cultural performances – and for the research that engages with such performances. Indeed, this relation is reflected in this volume itself. Its publisher Palgrave Macmillan (particularly through its two performance-oriented book series, *Studies in International Performance*, edited by Janelle Reinelt and Brian Singelton, and *Performance Interventions*, edited by Elaine Aston and Bryan Reynolds, in which this collection appears), along with Routledge, is currently one of the leading publishers of performance research, in part because it markets this research around the world. We, the editors, while critical of economic globalization, clearly do not imagine ourselves outside of its operation.

A second aim of *Contesting Performance* is to analyze both the global impact and decentering of performance research, especially given the United States' so-called 'imperial' present (even as that 'present' seems undermined by former US president Bush's disastrous foreign-policy initiatives during his administration) and the widely held view that performance studies is centered there. In the following we use the term 'performance research' rather than 'performance studies' to mark this decentering. Performance as a field of study first became institutional-ized in the United States, but has since gone global, a process that sounds uncannily like developments in economic and political neo-imperialism. But while cultural knowledge production has been decentered, it has not displaced what might be called the dispersed global West (Wee, 2007: 17–19), a West extended in part via humanities and social sci-ence research conducted in British and, particularly, US universities (Appadurai, 2000: 3). Thus it is evident that while familiar models of knowledge production that rest on binaries of center/periphery and inside/outside no longer easily apply, old power relations based on eco-nomics and politics have not simply or entirely been replaced by more diffuse cultural forces. To complicate matters still further, one cannot eas-ily separate 'good' from 'bad' political and socio-cultural 'influences' – in fact, the problematic borders between such influences are what need to be better understood. How, then, have scholars around the world variously incorporated, appropriated, decentered, and challenged a 'Western' or 'US' model of performance research for their own ends, anti-imperialist and otherwise?

The third major aim of this collection is to produce a better sense of the contours of performance research – and its stakes – by foregrounding its local contexts and trying to highlight 'other' voices and bodies of inquiry. The 'local' here refers not only to non-metropolitan locations, but also to sub-national or regional locations. To foreground the local, however, is not to argue that the local is more 'real' or privileged than other contexts, but rather to emphasize the local as a distinct context within which the globalization of Western power and knowledge is mediated, resisted, or appropriated. Indeed, the production of locality itself often occurs within a site traversed not only by the (dominant) US version of the global West, but also by regional powers and cultures. As shown by the chapters in the volume, whether it be Slovenia, Singapore, or Morocco, the local may contest, negotiate, and collude not only or even primarily with the global West, but also with more complex and multiple power relations.

In pursuit of these three aims, we have sought accounts of how per-formance research has emerged in specific places and solicited chapters from scholars around the world. Among the questions we posed to them: What defines the field, and what types of performances are studied? What constitutes 'proper' research, and which methods and theoret-ical frameworks have been important? Which researchers have been especially influential? Have there been particular performances, artists, genres, or practices that have helped shape the research agenda, and have any served as models for the very definition of performance? What problems have arisen around language and/or translation – in particular, how important or irrelevant is the English-language term 'performance' itself? What is the relation between 'practices' and 'models,' 'performance' and 'research,' 'performance' and 'teaching,' and how well do these distinctions translate or fit into local histories and situations? What critical questions or issues have been crucial, either in performance or research, or both? What institutions have supported and informed research, whether they be universities, governments, funding agencies, or other organizations? What types of curricula or pedagogies have been initiated? What publications have been important for researchers and practitioners? How has the research been shaped by broader social, political, and historical phenomena – for instance, by nationalism, war, genocide, diaspora, or globalization? What regional or international alliances or organizations have been formed? What challenges lie ahead? This collection brings together their responses.

Is performance studies imperialist?[1]

At the center of this collection is the effort to identify alternative genealogies to a research paradigm long associated with New York University's (NYU) Department of Performance Studies. Widely known as the 'broad spectrum' approach, its objects of study range, most famously, from ritual to theatre, and include performance art, dance, folklore, and performative acts of everyday life (Schechner, 1988). This research paradigm was first formally institutionalized at NYU in 1980, when the Graduate Drama Department became renamed the Department of Performance Studies. The department now hosts one of the field's most respected journals, *The Drama Review*, or *TDR*, and more importantly, has over several decades produced influential research. As argued in Marvin Carlson's *Performance: A Critical Introduction* ([1996] 2004) and Richard Schechner's *Performance Studies: An Introduction* ([2002] 2006), the origins of this particular research paradigm reach back to the mid-twentieth century. The breadth of research methods and theoretical frames is as wide as the broad spectrum of subjects studied, being drawn from such fields as anthropology, art history, communication, dance history, history, linguistics, literary studies, philosophy, postcolonial studies, psychology, sociology, and theatre studies. The methods include critical race studies, deconstruction, feminism, Marxism, new historicism, phenomenology, psychoanalysis, queer theory, semiotics, and speech act theory. And yet, as Jon McKenzie argues in *Perform or Else: From Discipline to Performance* (2001), the NYU paradigm was not totally diffuse: in its formative years in the 1960s and 1970s, it privileged theatre as a formal model for 'seeing' the broad spectrum of, while also privileging, liminal rituals as a functional model for theorizing the potential for performances to produce social change. Later, during the so-called US 'culture wars' of the 1980s and early 1990s (Graff, 1993), performance art became a paradigmatic performance genre, one that meshed well with emerging theories of subject formation found in poststructuralism and cultural studies (McKenzie, 2001: 29–53).

As some of the chapters collected here suggest, the success of the performance studies paradigm cuts two ways. There is little doubt that it has produced ground-breaking research that has been published and translated around the world. The NYU program itself remains one of the premier postgraduate programs in the field. Further, some of its graduates have gone on to teach in and, in some cases, administer other academic performance studies programs, while other graduates have entered careers in the arts, media, and other professions. NYU's influence extends far beyond the United States, as many scholars and artists have gone to that university from overseas for their training and then returned to their native countries to work. And yet, this powerful set of influences has produced 'blowback,' a critical counter-force against what some feel is the paradigm's dominant status.

One instance where this critical counter-force came to the fore was at the 2004 PSi conference held in Singapore, titled 'Perform: State: Interrogate.' During two of its main plenary sessions, in particular, presenters and audience members alike interrogated the US dominance of performance studies, the relation of this dominance to the American 'global war on terror,' the problematic role played by the English language in performance research, ranging from scholarly publications to academic conferences, and the very term 'performance' itself. The location and timing of these criticisms were significant. The conference in Singapore, where the state plays an active role in cultural affairs, took place in June 2004, just months after the Abu Ghraib prison scandal broke – detainee abuse had been committed by some personnel of the US 372nd Military Police company. Crucial to the critical reassessment of performance studies was the presence of over 130 Asian artists, scholars, and students, whose perspectives on performance research may differ greatly from US and UK artists, scholars, and students, owing both to their different performance traditions and to the legacies of American and British imperialisms in the region. Significantly, this critical force carried over to the following PSi conferences at Brown University in the United States and Queen Mary, University of London, where discussions begun in

Singapore not only continued but also were linked with debates over performance and human rights.

Given the collective weight of such critiques, is performance studies therefore to be considered imperialist? The answer depends both on one's perspective – in particular, on one's relation to the imperialist legacies just cited – and on how one defines or understands 'performance studies' (or 'PS'). It may be useful to see criticisms of the PS paradigm in terms of what might be described as a 'nested structure.' At the center is 'NYU PS,' or, more specifically, the performance studies identified with Richard Schechner's 'broad spectrum' approach. While Schechner has long been criticized on imperialist grounds (Boal, 1970), one cannot simply or at least simplistically dismiss Schechner's work on such grounds, as he has consistently taken positions critical of US policies, both international and domestic. And although Schechner's broad spectrum approach has certainly been influential, it should be stressed that the NYU program cannot be reduced to Schechner's work, as other notable researchers teach there, including Diana Taylor, one of our contributors, whose own Latin American background and training link her as closely to the imperial 'periphery' as to the 'center.' PS courses taught at NYU are, if anything, geopolitically *anti-imperialist* and (it should be noted) Schechner's own approach is itself contested there. Nevertheless, the wide-ranging adoption of the broad spectrum approach, alongside NYU's increasing global visibility – with study-abroad programs in roughly a dozen countries, and even an Asian branch of its Tisch School that opened in Singapore in October 2007 – has produced, at the very least, a palpable perception of NYU PS's dominant presence in the field.

NYU is itself 'nested' in a larger formation – 'US PS' or more broadly US/North American performance studies. Just as NYU PS is not reducible to Richard Schechner, US PS is not reducible to New York University. As Shannon Jackson's chapter in this collection argues, another performance research tradition – one based in oral communication – has contributed much to the field. Institutionally, Northwestern University in Evanston, Illinois, in particular, has produced many influential researchers, and over the last 15 years, many other performance studies programs or departments have emerged in North America. Some of these new performance studies programs are located at major research universities such as Brown, Stanford, the University of California at Berkeley, the University of North Carolina at Chapel Hill, and the University of Texas at Austin. But again, the very prestige of US performance studies and the sheer quantity of publications and presentations, while in one respect laudatory, nevertheless contributes to the sense of an American dominance of performance studies.

The political and social aspects of this dominance may surprise some scholars in the United States, who have long been accustomed to thinking of the field of performance studies as both deeply critical of hegemonic power formations and as itself *marginal*, barely recognized institutionally, and always searching for new methods and subjects to address central political questions of cultural identity, power, and resistance. However, what might be overlooked is the cultural specificity of US notions of cultural identity, power, and resistance when they are packaged as a seemingly universal form or format for multiculturalism. It should be noted that other fields, such as American and English literary and cultural studies, face similar dilemmas as their research and pedagogical approaches have also gone global (Gunn, 2001; Chen, 1998; and Pease 2002).

What further needs to be brought in here is a general thrust toward the marketization and commodification of higher education in parts of the world, in which the desire to have competitive development has meant that US research universities have become models through which 'innovation' and 'creativity' can be fostered for the 'new competencies' thought to be essential if less-advanced societies are to become knowledge nodes in the global circulation and production of information. Given this development, even apparently 'contestatory' fields of study such as performance studies or, for that matter, gender studies or postcolonial theory, can become suitable and commodifiable fodder for capitalist development, as their content becomes less important than their being part of the curricula for 'creative' tertiary education (Delbanco, 2007; Wee, in press;). For instance, we see in Uchino Tadashi and Takahashi Yuichiro's chapter on Japan how if, on the one hand, a specific national-cultural development trajectory has made it hard for performance studies to gain academic legitimacy, on the other hand, a certain performance studies model has become a pedagogical tool through which individuals can make better socio-economic impressions on business audiences – that is to say, performance studies has been 'embraced' by Japan's service-oriented, late-capitalist market economy.

Moving outward once again in the 'nested' structure of the imperialism of performance studies, we can focus on the *combined* American and British dominance of the field: 'US/UK PS.' As in the United States, the number of performance researchers and courses of academic study in the United Kingdom has increased in recent years, a development Heike Roms addresses in her chapter. Also important here has been the role of the Centre for Performance Research and of the journal *Performance Research*. As significant research has emerged from the United Kingdom, British scholars often explicitly contrast their research and pedagogy to that found in the United States, stressing in particular their emphasis on practice. However, for many scholars outside the United States and the United Kingdom, such distinctions are less important than these countries' perceived dominance, taken corporately as a kind of Anglo-American 'axis.' The leading publishers of performance scholarship – Routledge, the University of Michigan Press, and now Palgrave Macmillan – are based in these two countries. Their books are sold around the world, but marketed with an eye to US and UK readers. Further, of the 15 PSi conferences held thus far (1995–2009), two have been in the United Kingdom, and eight in the United States. There have been only five other host countries: Germany, New Zealand, Singapore, Denmark, and Croatia. Like the United States, New Zealand and Singapore are former British colonies, so from another perspective, only three PSi conferences to date have been held outside the Anglo-American sphere.

This point leads us to the last, but most vast, realm of the projected 'PS empire.' At the nested structure's outermost ring, we find 'Anglophone PS': the role of English as the lingua franca of performance studies. This dimension becomes clearer if, to the US and UK programs, one adds the study and research undertaken in Australia, New Zealand, Singapore, Nigeria, Kenya, Hong Kong, South Africa, and English-speaking Canada. At this juncture, though, we should note that performance research, broadly taken, follows a more general pattern. English has become the 'world language' in many fields of knowledge, as well as in the realms of international trade, finance, and transportation. But given the critical function of much cultural performance research, the global hegemony of English cannot be ignored.

The matter of translation arose at some PSi conferences; both the 1999 Aberystwyth and 2001 Mainz PSi conferences provided multilingual translations, while one plenary of PSi in Singapore was devoted to the topic of the power of English vis-à-vis other languages, especially those in Southeast Asia. However, at both Mainz and Aberystwyth, only a few plenary sessions were translated, while all the others were held in English. The entire discussion on language in Singapore occurred – not surprisingly – in English. On the one hand, these cases reveal that English provides a way for people whose primary language is not

English to communicate with someone for whom it is, and it also allows non-primary English speakers to communicate with one another. On the other hand, as was discussed at the Singapore PSi session, the predominance of English informs and deforms the concept of 'performance' and, by extension, the very phenomena studied 'as' performance. Several questions arise: How is this term 'performance' translated? When and why is it frequently left untranslated? And how do 'performance' and its translations resonate with other terms and usages? Lada Čale Feldman and Marin Blažević address issues of translation from a Croatian perspective in their chapter here.

Despite our interest in presenting alternative genealogies of performance research, this collection on performance research can itself be criticized for contributing to this 'PS empire.' One highly regarded scholar declined our invitation to contribute an essay precisely on these grounds, arguing that our attempt to survey the global growth of performance research would very likely extend the hegemony of the dominant PS paradigm, rather than truly offering and exploring alternative approaches. The editors recognize this risk. This collection is published in English by a major British academic press; we all work within the Anglophone network – Wee in Singapore, Roms in the United Kingdom, and McKenzie in the United States – and one of us not only trained at NYU but studied under Richard Schechner and sees him as a mentor.

Yet we believe the risk entailed is worth taking, for we strongly sense that the time has come for more open discussion and debate. Our sense of the collection's timeliness is grounded in a series of events that we have organized over several years, including a workgroup organized by McKenzie and Jessica Chalmers at the 2003 Performance Studies Preconference of the Association of Theatre in Higher Education; a three-day working session organized by McKenzie, Roms, and Wee at the 2004 Singapore PSi meeting; and a seminar organized by Roms and McKenzie at the 2006 American Society for Theatre Research conference. More recently, *TDR* responded to our question of performance studies and imperialism with a series of commentaries.[2] As we learned most keenly from the 25 participants at the Singapore workshop, many scholars feel a strong desire to 'tell their stories,' to relate local histories of the events and people, as well as the methodological tools and institutional challenges that inform and shape performance research. There is also great interest in simply learning what others are doing elsewhere. A critical mass has emerged that needs representation and analysis, despite the risk of expanding the alleged PS empire.

The West and the Rest?

But perhaps there is another danger lurking here. By posing the question of a PS empire in this Introduction, might we be reasserting and consolidating the oppressive notion of center/periphery – 'the West and the Rest' – that has long informed cultural and humanistic research? The foundational logic of historicism, in which modernity and modern research trends are seen to emanate from the center and then proceed to the world's 'hinterlands' (Sakai, 2006, pp. 170, 174), is hard to escape, despite the cultural-critical arguments made in the past 20 years or so regarding postcolonial hybridity and resistance, the colonial contact zone, alternative modernities, and so forth.

We believe that the chapters in this collection, taken together, indicate some form of an alternative. What emerges strongly from the collection as a whole is that the research – and indeed cultural and intellectual relationships between many of the non-American and non-Western European writers here – cannot be reduced to a dichotomous relationship between the globalized metropolitan West and the non-Western writers' specific localities. What the chapters of this book reveal is some displacement of the 'West and the Rest' paradigm, where the 'West' refers to NYU PS or, more generally, to the agendas of humanistic Anglo-American scholarship.

Certainly, in many respects, researchers around the world have to contend with the research agendas set in the 'center.' As prestigious UK and US academic journals and university presses have circulated critical and cultural theory dealing with issues of political transgression – such as race, gender, and postcolonialism – around the world, these models now shape non-Western performance research. But these paradigms can also overlook significant trends in cultural production and the production of cultural knowledge best understood from a local or regional perspective. For example, the term 'postcolonial' does not precisely describe the particularities of, say, post-apartheid or current Australian artistic critique. Edward Scheer and Peter Eckersall observe in their discussion of Australian PS's constant need to refer back to England and the United States as points of origin for cultural practices that there is a 'broad failure thus far to engage meaningfully with indigenous performance forms.' Loren Kruger's chapter on post-apartheid South African theatre argues that dominant performance studies models that stress the transgressive character of live performance miss the subversive local uses of commodified media forms.

The point here is that some research locations, though highly influenced by cultural-research developments in the 'advanced' center, also manifest distinctive national concerns and challenges. Sal Murgiyanto's chapter, for example, looks at collaborative performances within and between different Indonesian dance traditions as a way of affirming national, traditional, and contemporary art forms. Research in other locations manifests a strong set of relationships both to the West and to their immediate region. Bojana Kunst's chapter stresses the collaboration among performance researchers in Ljubljana, Zagreb, and Belgrade as a way to counter Western European and American understandings of regional cultural production. The researchers' voices in this collection do not always just talk back to the center, but speak for and among themselves, or to regional concerns and identities.

The importance of regions and regionalization have come to the fore in recent years for political and economic reasons (Katzenstein, 2005; Pempel, 2005). Regions have entrenched histories, with their own interior struggles between centers and peripheries that individual researchers often want and need to address. East Asians, for instance, need to deal with a triadic 'Japan-Asia-West' imaginary, rather than a simple West-Rest binary (Sun, 2007). In another example, the relationship within the European Union between its newer and more established members forms a complex set of intra-regional relationships that do not fit easily into a West-Rest paradigm.

Fundamentally, what many chapters in *Contesting Performance* indicate is that while performance research does emanate from an intellectual center in the Anglo-American West, and while the various culturalist agendas of that center have been globalized, such concerns are not always central shaping forces. Thus, the concerns of performance research scholars working around the world are not *inevitably* focused on disputing the global center's hegemonic status. While this does not mean that this volume collectively escapes the West-Rest or center-periphery set of binary oppositions, it at the least does suggest that the goal may be less to eliminate these binaries altogether than to *multiply* and *complicate* them in order to reveal a more complex analytic field.

Global sites of research

In an effort to give a vivid and nuanced sense of the ways that performance researchers around the world are grappling with problems of globalization, US hegemony, and the institutionalization of performance studies as a discipline, we are convinced that it is crucial to stress the 'situatedness' of local sites of research. It is in the dense particularity of specific places and times that the variety and subtlety of contemporary performance research emerges.

The collection is divided into three parts. In the first section, we track the experiences of scholars who tell different stories about the contextually situated institutionalization of performance studies. All narrate struggles over problems of language, disciplinarity, and academic institutionalization, though in intriguingly different ways. The second section underscores the importance of performance practice as a challenge to institutionalized understandings of performance, and suggests that it would be a mistake for academics not to take account of the ways that distinctive, local practices shape the possibilities of thinking about performance. The final section develops this point by bringing together scholars who argue that artists actually do the kind of productive and critical thinking about performance that we usually associate with critics and academic writing. The last chapters hence focus attention on artists who generate provocative, searching, and innovative performances, suggesting that through embodied, particular performance experience we may grasp genuinely distinct approaches to performance.

Part I   Institutionalizing performance studies

Our first group of chapters relates stories of successes in terms of where the academic discipline of performance studies has been planted and taken root in a divergent range of venues.

Diana Taylor in Chapter 1 traces the founding of the Hemispheric Institute of Performance and Politics, 'a collaborative, multilingual, and interdisciplinary consortium of institutions, artists, scholars, and activists throughout the Americas. Working at the intersection of scholarship, artistic expression, and politics, the organization explores embodied practice – performance – as a vehicle for the creation of new meaning and the transmission of cultural values, memory, and identity.' Transnational and trilingual, the Institute grew out of a desire to bring together archives of performance practice that had been far-flung and elusive. Deliberately refusing to call itself a 'center,' the Institute embraced 'a decentering project' that would attract scholars and artists from Canada to Patagonia, including 'scholars who do not usually consider themselves part of the hemisphere.' Cultural formations in the Americas themselves are often deeply transnational: 'Youth gangs span the Americas, transmitted by the Latino youth who grew up in the United States and deported for immigration reasons to a "home" in Latin America they never knew.' Confronting challenges of language, translation, limited media, and unequal distributions of power that sometimes threatened the notion of transnational partnership, the Institute has done well, with over 25 member universities and cultural institutions, thousands of participants at their '*Encuentros*,' a physical archive, and a huge online repository of digital materials, including a digital video library with 500 hours of streaming video of performance work. Taylor thinks of the Institute not as 'a thing' but as a performative 'practice.'

Gay McAuley takes us in Chapter 2 to the University of Sydney, where she narrates the history of the institutionalization of performance studies as a discipline. Neither isolated nor derivative, McAuley works to show how 'the Sydney-based scholars engaged with work being done elsewhere as they developed their own distinctive approach to the emerging discipline.' What emerged included a focus on regional performance practices, especially those in Indonesia and Japan, an insistence that students engage 'with contemporary performance practice via performance analysis and observation of rehearsal process,' and an interest in 'traditional text/narrative/character-based theatre' alongside non-theatrical and experimental performance practices. While optimistic about the fact that performance studies continues to grow in recognition in the Australian context, McAuley nevertheless foresees no escape from 'a field dominated by northern hemisphere institutions and perspectives, reinforced by a market economy within which the Australian experience is so often seen as utterly marginal.'

In Chapter 3, Heike Roms identifies two features that characterize performance studies in the United Kingdom: 'firstly, its attention to artistic practice, whether manifest as a renewed focus on the materiality of theatre or as the expansion of creative modes of investigation; and secondly, the tension between its close association with and frequent deliberate dissociation from its US-counterpart.' Contrasting with US approaches, Roms locates a possible 'British' concern with the study of performance within the recognition of creative practice as a form of research that possesses validity in the university. She proposes that '[w]hat may appear at first as a reversal of performance studies' focus from an extended consideration of cultural practices back to a narrow notion of performance as aesthetic production, in fact presents a profound re-evaluation of the nature of practice itself and our study of it.' Her chapter outlines what is at stake – institutionally and epistemologically – in such a re-evaluation for the researching and teaching of performance in the United Kingdom.

If performance research in such diverse sites as the Hemispheric Institute, Sydney University and the United Kingdom have flourished in part by contesting and reworking the dominance of US performance studies, Chapter 4 argues that even US disciplines and universities have had to resist and elude the oppressive techno-bureaucratic processes of globalization and of 'Americanization' itself. Shannon Jackson tells a story about the institutionalization of the 'other' tradition of performance studies in the United States – the oral, rhetorical tradition that came to be most famously housed at Northwestern University. Led in large part by Dwight Conquergood, the oral interpretative strand of performance studies stressed 'a particular kind of conversation between ethnography and performance – one about narratological politics and about cross-media translation' – and led to new methods in the field. The institutionalization process, though, did not mean that a contestatory and non-conformist edge was lost in pedagogy because of the pressures of techno-bureaucratic accounting and the pressures of the bottom line: 'performance pedagogy [...] is also terribly inefficient, requiring enrollment limits that do not make financial sense, requiring extended hours that challenge the classroom schedulers. [...] Indeed, it is a brand of performance that refuses to be measured by the system of inputs and outputs that structure the "performance evaluations" of academic departments with increasing frequency.' The rhetorical tradition of performance studies continues to be 'a relentlessly illegitimate, if undernoticed, discipline.'

The final chapter by Uchino Tadashi and Takahashi Yuichiro draws our attention to an unexpected institutionalization of performance studies, alongside a simultaneous deep resistance to performance studies as an import from abroad. Since Japan is often cited as a site where performance art itself originated, and since critical theories about performativity – for example, the work of Judith Butler and Eve Kosofsky Sedgwick – have been popular among Japanese academics, why, they ask, are there no performance studies departments at all, and only 'a handful' of Japanese books on performance art, experimental theatre, and cultural performance? Uchino and Takahashi point to long traditions of thought and institutionalization that led to a strict divide between academics – who study 'fixed' objects such as literary texts from empirical and historical perspectives – and journalists – who have cornered the market on contemporary, ongoing, and influential cultural experience. If performance studies has sometimes broken down such divisions elsewhere, it has not done so in Japan. Chapter 5 offers the instructive story of one academic, Sato Ayako, who was trained in NYU's Department of Performance Studies and returned to Japan to popularize the idea of 'performance-*gaku*' (performance 'scholarship' or 'discipline') as one embodied illustration of the situation analyzed. Sato deliberately turned away from the liminal and subversive model of performance, which she took to be characteristically 'American,' and instead formed her own model which stressed the effective playing of social roles. Uchino and Takahashi conclude that Sato's influential 'performance-*gaku*' is 'prescriptive,' reinforcing nationalist, class, and gender norms. With the translation from NYU Performance Studies to Sato's performance-*gaku*, then, came a profound transformation of methods and goals that has alienated progressive academics in Japan.

Part II Contesting the academic discipline through performance

This part contains chapters that suggest that performance practices and institutions *outside* of the university can productively drive thinking about performance. The contributors to this section discuss approaches to performance that are not strictly disciplinary or academic. Typically linked to the university's study of performance but inventive, unruly, and conceptual in their own ways, the chapters gesture to strong links between creative practice and critical reflection which have the capacity to transform performance research both within and beyond the university.

Edward Scheer and Peter Eckersall in Chapter 6 point to the value of experimental performance practice in disrupting the entrenched binaries that have shaped the adoption of US and European models of performance studies in postcolonial sites. Their context is Australia, where US hegemony and Australia's marginalization as the Antipodes – perpetually understood through its 'geographical oppositeness' – reinforce an oppressively colonizing sense that scholarship on performance always takes place in relation to Europe and the United States. Scheer and Eckersall favor moving beyond antipodality to a specifically regional understanding of Australian culture – reaching out to Japanese and Javanese performance practices, for example, to generate a sense of 'Australias beyond the antipodes.' They describe one performance, *Journey to Con-fusion*, as a crucial experiment in such regional collaboration, joining the Melbourne-based experimental performance group Not Yet It's Difficult (NYID) and Gekidan Kaitaisha (Theatre of Deconstruction) from Tokyo. Such performances embrace intercultural regional collaboration and the 'art of improvisation' (Meaghan Morris) as a 'usefully disruptive' approach to the antipodean logic of Australian performance studies as it is often conceived and practiced in the university.

In Chapter 7 Bojana Kunst discusses the Slovenian magazine *Maska*, which reflects in critical and in interdisciplinary ways on the question of performance through an ongoing engagement with 'the live practice of art.' *Maska* emerged as a dialogue between art and theory, and also as a give-and-take between local and global contexts. It created international visibility for Slovenian artists in the 1990s, who were developing new ideas about performance, but it also introduced theoretical approaches from around the world to a Slovenian audience, and broadened the conversation about performance through translations and visiting lecturers. Kunst argues that *Maska* can also help to reorient global understandings of Slovenian art. Widely misunderstood in Western Europe and the United States, Slovenian approaches to performance in the 1980s and 1990s did not always adopt 'dissident' attitudes; nor did they all strive to catch up with capitalist Western Europe, as the conventional teleological story about the transition from communism to capitalism assumes: *Maska*' has always tried to think the processes of art through more complex connections and has problematized the mirrors that we hold in front of each other during these processes.' The magazine is intended to act as a complex and dynamic 'platform for the production of knowledge' that works outside of academic institutions while drawing on ideas from globalized performance studies and from local artistic practices.

If Australian artists have reconceived their practices in relation to regional rather than colonial powers, and Slovenian artists and thinkers have reacted in a complex set of ways to the collapse of communism and the impact of commercialization and globalization, academics and performance artists in the city-state of Singapore face yet another set of political circumstances: a highly centralized state which issues powerful performative declarations and uses carefully staged public performances – such as National Day Parades – to shore up its own power. In this context, according to Ray Langenbach and Paul Rae in Chapter 8, none of the institutionalized approaches to performance – 'whether internationally produced theoretical perspectives, local research commissions, or "objective" academic scholarship; whether "thick" or "thin" analyses – has yet proved capable of modeling an appropriate relationship between cultural performances and their researchers [...].' The authors pick out a number of fruitful new directions in scholarship, such as densely researched studies of local traditions – for example, Margaret Chan's *Ritual is Theatre, Theatre is Ritual* (2006) – and attempts to contextualize Singaporean performance in a regional context – for example, Jennifer Lindsay's edited *Between Tongues: Translation and/of/in Asia* (2006). But they place their highest hopes in ongoing performance practices and artistic networks, such as a roundtable called 'Panic Buttons: Crisis, Performance, Rights,' where artists, activists, and academics from Singapore, Malaysia, Indonesia, and Thailand developed an ongoing and dynamic network for dialogue and exchange that would have been impossible, according to Langenbach and Rae, if located wholly within the university setting. Privileging such regional 'hybrid events,' they call for a wide, regionally based understanding of the field of performance research that reaches beyond the walls of the academy.

Sybille Peters then turns our attention in Chapter 9 to the German context for performance research, which, she suggests, is defined neither within the dominant Anglo-American institutionalization of performance studies nor in opposition to it. Here, Peters argues, the future of performance research relies on the development of hybrid formats 'that allow transitions between research on performance and performance-as-research' and thus challenge the strict separation between theory and practice that has defined the German scholarly tradition. Like Langenbach and Rae, Peters locates such hybrid formats both within and outside of the academy. She offers an overview of a variety of different approaches, from well-funded, university-based interdisciplinary research programs to independently curated and produced interventionist performance projects. These approaches have the potential to 'lead to a different understanding of research, which regards it no longer as the privilege of the specialist fields of science, theory, and art. Instead, the role of these fields would be to help organize and make visible the collective research that is undertaken by everybody, every day, making use of a wide range of procedures and integrating all forms of knowledge.' By thus questioning the dividing lines that currently separate the institution of the university from art and politics, such non-specialized hybrid research, Peters proposes, may ultimately lead to a form of egalitarian politics in our knowledge society. But she also expresses doubt over the future likelihood of such transitional performance research – in a time of increasing economic pressure on academic institutions, ' "performance",' she remarks succinctly, 'may simply come to stand for "output" and "research" for what economists do when they analyze markets.' This is a fear that has resonance with Uchino and Takahashi's concerns with the commodification of performance research and pedagogy.

In the last chapter of this section, the difficulties of translating not only the practices of performance studies but the English term 'performance' itself become clear in Chapter 10 about Croatia by Lada Čale Feldman and Marin Blažević. Academics in Croatia work under the constant pressure to translate texts and traditions from elsewhere, and what Anglophone scholars call performance in Croatia joins together a complex amalgam of many cultural and intellectual traditions, some in conflict with others: German historically oriented traditions of *Literatur-* and *Theaterwissenschaft* and *Volkskunde*, Russian formalism and Prague structuralism; French structuralism, narratology, and the Tartu school; Eco's semiotics and theatre studies; and the translation of excerpts from such works as Schechner's *Performance Theory* (1988), Butler's *Gender Trouble* (1989), and Turner's *From Ritual to Theatre* (1982), among others. In this complex, multinational context, Feldman and Blažević argue that contested terms and multiple conceptual traditions characterized Croatian scholarship on theatre, folklore, folk drama, and performance art for many decades. In the mid-1990s the magazine *Frakcija* 'introduced a comprehensive generic term – "performing arts" (*izvedbene umjetnosti*) – to bring theatre and performance art closer to each other within the sphere of theory and criticism.' While this concept of performance has broken down traditional boundaries and brought together an array of practices under the umbrella of 'performance,' Feldman and Blažević too stress that it is aesthetic performance practices rather than academic criticism that have offered the contestatory reflections and done most to disrupt old binaries and to generate a lively Croatian culture of 'performance.'

Part III   The power of performance practice

If some of the previous chapters mark how scholars can embrace local artistic practice as 'a material practice of thinking' (Kunst) that connects with the academic discipline of performance studies in ways that may reinscribe it, the final section offers chapters on how artists themselves may offer productive thinking about performance in specific sites. The scholars in this section study artists whose performances help us to grasp and think through the political, social, and cultural contours of their local contexts, suggesting that artistic performance *itself* may resist and help reconfigure global performance research.

Khalid Amine argues in Chapter 11 that Moroccan theatre has often oscillated between an embrace of Western proscenium theatrical models and a fierce rejection of European paradigms in the name of a return to 'indigenous' performance traditions. Contemporary experimental theatre practices, according to Amine, explore a 'liminal space' that fuses Western theatrical traditions and local Arabic performance traditions. Deliberately hybrid and self-reflexive, Moroccan artists draw on long local traditions, such as the contemporary street performances known as *Al-halqa*, subversive comic theatre, and improvisational performance, to disrupt the hegemonic practices of politics and official cultural institutions, while also contesting legacies of colonial rule. And it is in these hybrid performances that 'liminal third spaces' emerge, spaces 'that transform, renew, and recreate different kinds of writing.'

Indonesian dance offers some of the same dynamic theoretical possibilities as Moroccan theatrical practice, according to dance scholar Sal Murgiyanto. Indonesian artists deliberately fuse innovative, contemporary practices with a vast array of 'traditional' dance forms from different ethnic traditions in order to contest both colonial cultural power and any single, official version of Indonesian culture. Dance, in Murgiyanto's account in Chapter 12, begins to erase the boundaries between performance and performance research, since it can be considered in itself a species of 'research.' First, by privileging collaboration among ethnic groups and traditions, Indonesian dance becomes 'a significant means for "thinking" and "imagining" a contemporary dance' that draws on both 'established "national traditional dances" and "classical dance" training.' Secondly, performance practitioners here, as in Morocco, develop their resistance to colonial legacies by deliberately exploring a heterogeneous range of traditional forms to make contemporary art works. Contemporary Indonesian dance, then, offers not only a body of performances to study but itself constitutes an 'approach' to performance, fusing complex national, traditional, and contemporary strands 'in a practical act of research.'

In another national context, Sharon Aronson-Lehavi and Freddie Rokem urge us in Chapter 13 to see contemporary Israeli theatre as a productive response to Israeli politics, with its vexed and double identity as both oppressor and oppressed. They claim that the transformation from Hebrew as a language of religious practice to an official language of state power has been '*the* most significant performative act of Israeli culture,' and argue that Israeli performance is now playing a crucial role in unsettling the hegemonic power of Hebrew in constructing a monolingual Israeli identity. One example is Ilan Ronen's polylingual production of Beckett's *Waiting for Godot* (1952): 'When the performance played before Arabic audiences, Didi and Gogo spoke in Arabic, and when the play was performed for predominantly Jewish audiences, they spoke in Hebrew with a marked Arabic accent. Pozzo always spoke in Hebrew, while Lucky spoke in a distorted Classic Arabic.' Israeli playwrights and performance artists have insisted not only on multiple languages but have also adopted critical performance paradigms from around the world, including feminist and postcolonial critiques, and so 'have gradually enabled non-hegemonic bodies to voice themselves [...], deconstruct[ing] the relations between word and action and enabl[ing] the re-examination of Israeli identity by exploring the performer's culturally multi-layered body.' Aronson-Lehavi and Aronson conclude by praising the increasing reliance on the human body in Israeli performance: 'The body has a different presence than words, since it is the action itself, not an additional layer of commentary and reflection.' And it is the specificity of this located performative embodiment, they suggest, rather than abstract commentary and critique, that could help to usher in a better future.

Concluding this section, Loren Kruger in Chapter 14 calls attention to another 'failure' of performance studies' translation into a new context – in this case, post-apartheid South Africa – by arguing for the capacity for multimedia and commodified 'performances' to be subversive. While the live stage of Johannesburg's Market Theatre once provided a safe haven for anti-apartheid protest drama, South African theatre now competes in a profoundly transformed cultural landscape with the performance sites of a profit-driven leisure industry on the one hand, and the well-funded products of television and other multimedia forms on the other. Criticizing the 'theatrical exceptionalism' that has dominated theatre scholarship as well as performance studies approaches, which privileges live performances as the most authentic forms of resistance, Kruger urges a different, more mediated model that relies on a Brechtian notion of 'stealth.' Rejecting the 'heroic realism' that marks traditional protest theatre in favor of a 'critique by cunning,' she proposes an approach that reveals the subversions of commodified and multimedia art forms. Kruger asks performance researchers to attend to 'the cunning mode of critique under cover of commodity production which might characterize public/private museum practice that encourages tourist consumption as well as edification, or of television series that address viewers as agents of their own education *and* as consumers of sponsors' products.'

As this survey of the chapters in *Contesting Performance* suggests, the breadth and contrast of perspectives found across contemporary performance research caution us against holding an overarching notion of exactly what this research 'is,' even while also indicating that its institutionalization as – and in some cases *against* – performance studies will continue to take a variety of forms, forms shaped by a complex mix of local, regional, and global forces. Amid these forces – cultural, historical, political, economic, technological, environmental – performance will no doubt continue to contest socio-cultural injustice, even while its very concept and understanding remain contested, and even if the very notion of 'contesting performance' contains within it certain cultural and historical tensions that may not always and everywhere be relevant. This volume provides a global snapshot of performance research in the early twenty-first century and opens up new perspectives on the relation between performance and performance research.

Notes

1. An earlier and more extended version of a part of this section previously appeared in *The Drama Review* (see McKenzie, 2006). In another issue of *TDR* that addressed this question, Janelle Reinelt (2007) argues for the need to develop international performance literacies. For more responses to the question, 'Is performance studies imperialist?,' that appeared in *TDR*, including responses from Richard Schechner, Diana Taylor, and Takahashi Yuichiro, among others, and an editorial cartoon by Diana Raznovich staging the question '*And We are the Imperialists??????,*' see Raznovich et al., 2007.
2. See note 1.

References

Appadurai, A. 'Grassroots Globalization and the Research Imagination,' *Public Culture*, 12 (2000) 1–19.

Boal, A. 'Letter to the Editor,' *The Drama Review (TDR)*, 15:1 (1970) 152–4.

Butler, J. *Gender Trouble: Feminism and the Subversion of Identity* (London and New York: Routledge, 1989).

Carlson, M. *Performance: A Critical Introduction* (London and New York: Routledge, 1996; 2nd edn 2004).

Chan, M. *Ritual is Theatre, Theatre is Ritual: Tang-ki: Chinese Spirit Medium Worship* (Singapore: SNP Reference, 2006).

Chen, K.-H. (ed.) *Trajectories: Inter-Asia Cultural Studies* (London and New York: Routledge, 1998).

Delbanco, A. 'Academic Business: Has the Modern University Become Just Another Corporation?,' *New York Times Magazine* (30 September 2007).

Graff, G. *Beyond the Culture Wars: How Teaching the Conflicts can Revitalize American Education* (New York: W. W. Norton, 1993).

Gunn, G., (ed.) 'Special Topic: Globalizing Literary Studies,' *PMLA*, 116:1 (2001).

Katzenstein, P. J. (ed.) *A World of Regions: Asia and Europe in the American Imperium* (Ithaca, NY: Cornell University Press, 2005).

Lindsay, J. (ed.) *Between Tongues: Translation and/of/in Performance in Asia* (Singapore: Singapore University Press, 2006).

McKenzie, J. *Perform or Else: From Discipline to Performance* (London and New York: Routledge, 2001).

———. 'Is Performance Studies Imperialist?,' *The Drama Review (TDR)*, 50:4 (2006) 5–8.

Pease, D. E. (ed.) *The Futures of American Studies* (Durham, NC: Duke University Press, 2002).

Pempel, T. J. (ed.) *Remapping East Asia: The Construction of a Region* (Ithaca, NY: Cornell University Press, 2005).

Raznovich, D., R. Schechner, E. Barba, K. Tomoko, Y. Takahashi, W. H. Sun, D. Taylor, and G. Gómez-Peña. 'Is Performance Studies Imperialist? Part 3: A Forum,' *The Drama Review (TDR)*, 51:4 (2007) 7–23.

Reinelt, J. 'Is Performance Studies Imperialist? Part 2,' *The Drama Review (TDR)*, 51:3 (2007) 7–16.

Sakai, N. ' "You Asians": On the Historical Role of the West and Asia Binary,' *Japan After Japan: Social and Cultural Life from the Recessionary 1990s to the Present*, ed. T. Yoda and H. Harootunian (Durham, NC: Duke University Press, 2006), pp. 167–94.

Schechner, R. 'Performance Studies: The Broad Spectrum Approach,' *The Drama Review (TDR)*, 32:3 (1988) 4–6.

———. *Performance Studies: An Introduction* (London and New York: Routledge, 2002; 2nd edn 2006).

———. *Performance Theory* (London and New York: Routledge, 1988).

Sun, G. 'How Does Asia Mean?,' trans. S.-L. Hui and K. Lau, *Inter-Asia Cultural Studies Reader*, ed. K.-H. Chen and B. H. Chua (London and New York: Routledge, 2007), pp. 9–65.

Turner, V. W. *From Ritual to Theatre: The Human Seriousness of Play* (New York: Performing Arts Journal Publications, 1982).

Wee, C. J. W.-L. *The Asian Modern: Culture, Capitalist Development, Singapore* (Hong Kong: Hong Kong University Press, 2007).

———. 'Once Again, Reinventing Culture: Singapore and "Globalized" Education,' *Traces: A Multilingual Series of Cultural Theory and Translation*, in press.

Part I
Institutionalizing Performance Studies

1
The Many Lives of Performance: The Hemispheric Institute of Performance and Politics

Diana Taylor

The Hemispheric Institute of Performance and Politics is a collaborative, multilingual, and interdisciplinary consortium of institutions, artists, scholars, and activists throughout the Americas. Working at the intersection of scholarship, artistic expression, and politics, the organization explores embodied practice – performance – as a vehicle for the creation of new meaning and the transmission of cultural values, memory, and identity. Anchored in its geographical focus on the Americas (thus 'hemispheric') and in its three working languages (English, Spanish, and Portuguese), the Institute seeks to create spaces and opportunities for cross-cultural collaboration and interdisciplinary innovation among researchers and practitioners interested in the relationship between performance, politics, and social life in the hemisphere.

While its administration is housed at New York University, the Hemispheric Institute is comprised of over 25 member universities and cultural institutions throughout the Americas. Institute initiatives include courses, work groups, conference-festivals (*Encuentros*), a Digital Video Library (HIDVL), archives, an online scholarly journal (*e-misférica*), a trilingual website, an emerging performers program in New York City (EMERGENYC), and public online forums. In 2008, the Institute inaugurated the *Centro Hemisférico*, a collaborative research center and performance space in Chiapas, Mexico, in partnership with FOMMA (a Mayan women's theatre collective) as well as its Hemispheric New York center, which organizes public events that feature artistic and scholarly work produced in New York City.

(Hemispheric Institute of Performance and Politics, 2009)

The Hemispheric Institute, Hemi for short, was started in 1998 with support from Ford, Rockefeller, and, later, the Andrew Mellon foundations. This project, like many I suppose, began as a way to solve a few related problems. Anyone working in Latin American theatre and performance knows how difficult it is to access the materials that interest us. One day, sitting in my office in Performance Studies at NYU, a few Latin American alumni and doctoral students and I started thinking out loud. How could we continue our collaborative projects on performance in the Americas? The first life of performance, so vital for communities who have participated in some of the genre-shaking productions, eludes most of us – the early *actos* by El Teatro Campesino in California as part of the United Farm Worker's strike, feminist performance artists starting at WOW Café, Enrique Buenaventura's experimentation with T.E.C. in Colombia, Augusto Boal's development of *Theatre of the Oppressed* as part of a literacy campaign or his *Legislative Theatre* when he ran for political office in his native Brazil, Grupo Cultural Yuyachkani's work with Native Peruvian communities in the 1970s or with the Truth and Reconciliation Commission in 2000, Jesusa Rodríguez's constant critique of power structures in Mexico in the 1990s and 2000s. Even the scripted Latin American performance pieces rarely travel beyond their own cities, and most of the practices go unrecorded. You had to be there.

The second life of performance – the scripts, videos, newspaper accounts – also prove elusive. Performance and scholarly texts are seldom published and translated. Videos of performances prove still harder to locate. Developing a course syllabus often feels like starting from scratch. Small wonder there is little scholarship on Latin American theatre and performance. What performances have scholars seen? What archives can they draw from? Yet ironically, after working for 20 years in performance (broadly understood) in the Americas, several of my colleagues and I had individually amassed or identified collections of important, hard-to-find research materials on theatre, performance art, rituals, processions, and political demonstrations, and had no place to put them. Nor were they doing anybody any good. The materials were not published. They were not archived. And even if they had been, the scholars, artists, and students who might find them interesting would not have had access to them. Libraries in Latin America are woefully under-stocked; books rarely circulate from one country to another; language barriers complicate access; few people have money for professional travel to specialized archives and libraries, even if these had been collecting performance materials. When we started, online archives and libraries were still a dream. So what would constitute our object of analysis?

Performance research, understandably, then, was not a well-established area of concern in Latin America in the 1990s, although some notable work was being done. Scholarly journals such as *CONJUNTO*, published by the prestigious Casa de las Américas in Cuba, have made work by theatre and performance scholars and practitioners available for the past 40 years. Individual theatre and performance collectives have for decades offered performance workshops and seminars in their own countries. Since the 1960s, Grupo Cultural Yuyachkani has worked closely with most of the actors, directors, playwrights, and theorists in Peru. Miguel Rubio, their artistic director, has written books that circulate among theatre and performance practitioners in the Americas. Santiago García, artistic director of Teatro La Candelaria in Bogotá, Colombia, has done the same. Other pivotal figures have furthered thinking about performance theory and practice in their cities and countries throughout Latin America. Some national organizations, such as Mexico's National Arts Center, have included research and archival dimensions to their purview. The Centro Investigación Teatral Rodolfo Usigli (CITRU) has funded researchers and published excellent studies since it came into being in the 1980s. While it originally confined itself to theatre, it gradually opened up to research on dance, music, performance art, and eventually to some of the interdisciplinary scholarship scholars have come to identify with performance studies. But sustained, hemispheric-wide performance research hardly existed until 1989 when a group of practitioners, under the leadership of the eminent Argentine theatre director and playwright Osvaldo Dragún, started the Escuela Internacional de Teatro de America Latina y el Caribe (*EITALC*, the International Theatre School of Latin America and the Caribbean).[1] This workshop, also sponsored by Casa de las Américas, happens several times a year in different parts of Latin America and the Caribbean, bringing noted directors and actors together for anywhere from two weeks to a month to share performance work. As stated on EITALC's website, this group has three main functions: to develop and 'perfect' theatre work, to support research, and to produce publications and video materials. The emphasis, more artistic than scholarly, has helped strengthen a community of artists throughout the area (see EITALC, n.d.).

Our idea when we started the Hemispheric Institute, then, was to focus more on the research and methodologies aspect of performance work. We created a consortium of institutions, beginning with the home institutions of these NYU Performance Studies alumni – Zeca Legiero of UNIRIO (Universidade Federal do Estado do Rio de Janeiro, Brazil) and the Autonomous University of Nuevo León (Monterrey,

Mexico), where Javier Serna taught. We added institutions that house performance-oriented scholars and started working together to share our materials, teaching experience, and methodologies through the use of Internet technologies that we have developed as we went along. Soon, it became clear that we could do more. We could actively involve artists and activists in exploring the intersections of performance and politics by bringing us all together for an annual ten-day intensive workshop/conference/performance festival or *Encuentro*. We could create new knowledge and actually help build the field we were purportedly studying through our team-teach courses, our ten-day *Encuentro*, our e-journal *e-misférica*, and year-long working groups and projects we developed with artists and activists. From all of that, we could develop an important archive of performance research materials.

Although the activities can be enumerated quickly and simply, the process has been ongoing and far from simple. While the issues have changed as we have gone along, every aspect has been negotiated. The early discussions with the initial co-founders from Brazil, Mexico, Peru, and the United States included: Why Hemispheric rather than Americas? America is a highly contested space, and much blood has been spilt over who gets to claim and define it. The small, final 's' in Americas sounds more like an afterthought than a true bid for plurality and inclusivity. 'Hemispheric' has the advantage of blurriness – no one knows exactly what it means: Right lobe? Left lobe? Thus it discourages passionate identification.

The scope of the project encompassed the Americas – from Canada to Patagonia – as a space of shared histories and practices without limiting itself to this landmass. Many of our participants focus on the traveling performance practices produced by Diasporas, displacements, trajectories of migration, and globalization. Why Institute? We all shunned the notion of a 'center' in a decentering project. Languages? Tri-lingual to start – English, Spanish, and Portuguese, even though we recognize these are all colonial languages. Our platform? We agreed to start working in and through universities to capitalize on existing resources and build up a stable infrastructure that would allow us to expand and include non-academic participants. Why NYU as the administrative site? The initial founders had meet in the Department of Performance Studies. Jill Lane, then a Performance Studies doctoral candidate, worked as the first graduate assistant; Lucia Wright, a Performance Studies alumna, and her husband Randy Wright served as our first webmasters; and Julie Taylor, a doctoral candidate in Spanish and Portuguese, was our first Assistant Director. As important perhaps, I was willing to do the fund-raising.

We also wanted to attract US-based scholars, artists, and scholars who do not usually consider themselves part of the hemisphere. How to start? Simple (we thought): we would do what we normally did in universities, but do it together, that is do research and teach about performance, but do it collaboratively, across national, linguistic, and disciplinary divides.

The plan was straightforward. We started with the courses. We designed a chronological series of four courses on performance and politics starting with Conquest, then Colonialism, moving to the nineteenth century in 'Staging the Nation,' then on to the twentieth and twenty-first centuries in 'Globalization, Migration and the Public Sphere.' Performance, as a repertoire of embodied practice, was central to our investigation of historical processes, considered alongside traditional archives of documentary evidence. Our first course description in 1999 read: 'Performance was fundamental to both indigenous and European colonial epistemology, and was a primary means through which both cultures maintained or contested social authority. We analyze the profound changes wrought on these performance cultures in their encounter with the other, and examine how performance was strategically altered and used by various social groups in order to achieve their ends.' Having decided on the topic and focus, the faculty from our various universities planned the course together, breaking the materials into units (or modules), uploading the materials online, and deciding how much time each one of us wanted to dedicate to each unit in our individual classes yet give our graduate students a shared experience. Students work together online in a multi-lingual, international, and multidisciplinary setting and develop collaborate final projects – everything from simple webpages on performance practices to multi-media *web cuadernos*, including videos, audio files, photographs, interviews, bibliographies, and other materials – that flesh out their own scholarly interventions as they create their own archives.

This sounds easy now, but back in 1998 online course materials were rare, and few people outside First World institutions had easy access to the Internet. Almost none of our students in the United States or Latin America knew how to build their own websites. The drama of organizing student 'chats' and discussions online to develop projects in the days before Instant Messaging lies buried deep in the archaeological past of our site. Archiving performance videos online was considered impossible. And in some partner institutions, the interdisciplinarity required in performance studies posed a problem – the way courses were listed and approved made clear that collaborative research belonged in the sciences, but not in the arts and humanities.

Our first *Encuentro*, hosted by Universidade do Rio de Janeiro, a public university, focused on 'performance and politics,' but not in the way we imagined. The entire public university system of Brazil was on strike, and we were allowed to continue only because the *Encuentro* was an international event that had been planned a year in advance. Artists, activists, and scholars all started to discuss performance and politics as strikers camped in tents around our conference space, interrupting the performances with jeers and chants. Politics crashed up against performance in this first event as it has in all others – though the confrontations and contradictions play out differently from year to year.

Yet the *Encuentro* exceeded all expectations. The organization of the children of the disappeared from Argentina (HIJOS) found their counterparts in Brazil, and together we organized a protest *acción* (act or intervention), starting in an infamous detention and torture center in downtown Rio. With Grupo Arte Callejero, a collective of graphic artists who work with HIJOS and other human rights groups, we created road signs that we hanged along the main thoroughfare of Rio with the dates of all the military coups in Latin America: Brazil 1964, Chile 1973, Argentina 1976, and so on, along with signs for School of the Americas and the IMF to link US military and economic practices to Latin American atrocities.[2] In a more productive show of international collaboration, the Brazilian HIJOS protected foreign nationals by hanging the signs themselves while we all helped and supported them. Augusto Boal offered a workshop on Theatre of the Oppressed methodology that taught scholars – Doris Sommer speaks repeatedly of this – how we learn from *doing*, not just observing. Grupo Cultural Yuyachkani presented *Antigona*, a solo-piece developed by Teresa Ralli, the solo performer and a founding member of the group, reflecting her work with survivors and witnesses of the civil war in Peru. What happens when we witness atrocity but cannot act to protest or prevent it? Ismene, the narrator of this *Antigona*, offers hope for those who could not act heroically in the face of imminent death, and who responded only belatedly to the trauma of their country. Theory, art, activism – it is hard to tell them apart in projects in which every artistic practice puts forward a theory, and every theory needs to be put into practice.

The theory-art-activism overlaps, like much else we discovered, need to be put into action rather than talked about. Little by little we developed and formalized the concept of working groups. During the following *Encuentros* ('Memory, Atrocity, and Resistance,' Monterrey, Mexico 2001; 'Globalization, Migration, and the Public Sphere,' Lima, Peru 2002; 'Spectacles of Religiosities,' New York, USA 2003; 'Performing Heritage,'

Belo Horizonte, Brazil 2005; 'Corpolíticas/Body Politics: Formations of Race, Class and Gender in the Americas,' Buenos Aires, Argentina 2007) working groups of artists, activists, and scholars became the backbone of the event. People propose or sign up for a working group that meets for two hours daily throughout the event and presents something on the final day. Each group decides what it wants to do and how to organize itself. Some meet again each *Encuentro*; others work year round on special projects either online or through small working group meetings. The work group focused on Intangible Heritage was asked by UNESCO to write the 'Manual for the Implementation of the International Convention on Intangible Cultural Heritage' in 2005. Another working group is charged with developing a special collection on our website dedicated to the Native Americas, which brings together existing performance materials from our archives, courses, e-journal, *Encuentros*, and working groups, and augments them into a coherent and searchable collection available to all. There are several working group initiatives of this magnitude and importance.

In short, the project took on a life of its own – not always in the form we envisioned, but it worked. Now, ten years later, we have over 25 member universities and cultural institutions; thousands of people have participated in our courses and events. We have developed a major digital video library that includes 500 hours of streaming video of performance in the Americas, a physical archive, and a massive repository of digital materials that we are currently curating and making accessible to all viewers online. Our *web cuadernos* are now considered serious scholarly contributions to our field, and have been evaluated as part of tenure documents and distributed by national organizations as valuable teaching materials. Yet, it is clear to all of us that the project is ongoing – not a thing but a practice.

The challenges have been (and continue to be) formidable. They also change over time. At first, some of the problems had to do with definition, language, and power. 'Performance' – the word (and often the practice) – does not translate, as I have argued elsewhere (Taylor, 2002). In Latin America, the term referred almost exclusively to performance art, especially live art originating in the visual arts. Moreover, suspicions of imperial imposition attached themselves to the unwelcome loan word. If the word is foreign, some argued, the concept must be too. With time, the intensity of the debates about terminology dissipated, and participants began to appreciate the multi-layered term 'performance' that links artistic, social, economic, sexual, managerial, and other imperatives. 'Performance,' most now agree, exceeds live art by referring to acts without explicit aesthetic aspirations, such as rituals, political demonstrations, and gendered, raced, classed, and 'aged' performances of the self. 'Performance' also exceeds the word *acción* in Spanish, roughly translated as 'intervention,' which is often offered as a synonym. Not all performances are interventions. Some are so normative that they masquerade as transparent, and others may lack the consciousness of being an 'act.' Jon McKenzie's *Perform or Else* (McKenzie, 2001) makes clear that the corporate imperative to 'perform' demands that we account for economic factors and managerial style. The term 'performance' has gradually come to incorporate these many valences, and the prestigious publishing firm, Fondo de Cultura Económica, the largest Spanish-language press in the world, is launching the first books – Richard Schechner's *Performance Studies: An Introduction*, and *Theories of Performance*, edited by Diana Taylor and Marcela Fuentes – in what we hope will be a performance series. Theatre and cultural studies journals now publish articles on performance, broadly understood.

Yet translation has proved stressful in other ways. At our first *Encuentro* in Rio de Janeiro, Brazil in 2000, we did not have enough money to budget in translation. Any fantasies we had that we would somehow understand each other died quickly. We called on all those with language skills and generous personalities to help us through. During the wrap-up session at the close of the *Encuentro*, two fundamental things became clear. First: we needed professional tri-lingual translators – at least for the major talks and round-tables. Translation into native languages has been added when necessary. Second: we had to recognize the deeper truth that we do not know or understand each other in the Americas. Art is not the only thing that resists translation. We cannot translate everything, and there will be things we cannot grasp, no matter how well we translate. The price of admission, as Guillermo Gómez-Peña so compellingly put it during the wrap-up, is that we accept that we will always miss something. No one in the Americas enjoys a privileged position of having access to everything – so we agreed to start there, with that premise. Participants work to understand each other as much as possible, and they are creative and generous in finding ways to communicate.

Some of the problems arose from our work online, starting with the lack of basic infrastructure throughout much of Latin America, and the enormous disparities for faculty and students in terms of resources and access. One university had no Internet access when we started the project. These economic inequalities required that we rethink existing technologies and protocols to address our needs. While we welcomed the democratizing potential of technology's 'global' reach, and its ability to create new publics, we feared that unexamined terminology and practices created their own exclusions. Terms such as 'digital divide,' which signal concerns about access to new technologies, simultaneously threatened to reinscribe notions of underdevelopment in regard to knowledge production. Existing distance learning models lent an 'export' quality to knowledge – marketed in terms of 'acquisition' rather than 'production.' English, moreover, had become the lingua franca of the Internet, leaving huge populations out of important debates. Our goal was not to deliver materials to new audiences (one-way transaction), but to create a multi-lingual workspace that built capacity at all the sites. The principle guiding this project is one of partnership in producing knowledge, rather than simply delivery of existing courses online. We worked on the collaborative, interactive, and creative uses of technologies to further our collective work in performance and performance studies.

Interestingly, we faced another form of performance anxiety: we needed to resist the lure of faster and fancier technology. While many sites outperform ours, the countries we work with do not have the bandwidth or resources to download large files and acquire the latest software, which limits their access to much online material. Our website, while very extensive, was deliberately designed to be straightforward, fast, and easy to use in order to maximize its accessibility in places with old computers, slow connections, limited software, and novice users. True, the website also exhibits the mix-and-match quality of our different designers and media consultants over time, and it was only in 2006 that we decided to overhaul our entire site and move it into a searchable content management system and develop the interface with HIDVL (Hemi's digital video library). Until recently, all materials were uploaded according to Hemi events – courses, *Encuentros*, and so on. A user would need to know when an artist had performed to find a video of her work. Our new dynamic website, which was launched in 2009, invites visitors to view by topic, by special collection, or by specific Hemi events. Our innovations, then, lie in the ways we have creatively developed and adapted digital technologies for cross-cultural, multi-lingual, interdisciplinary use. When Hemi's work with HIDVL became story of the month on Apple's website in 2006, we felt that perhaps we still were at the cutting edge of performance and the Internet, even given our no frills interface.

Digital technologies, of course, raise many issues too about the many lives of performance. The ubiquitous/unlocatable nature of the virtual gives a whole new twist to discussions of archives and repertoires, enduring and ephemeral materiality, presence, distance, and time. Like traditional archives, digital web archives seem built to last. I personally am heavily invested in that supposition. The Hemispheric Institute works hard to curate, catalogue, preserve, and upload materials. Yet, I have never experienced ephemerality to the degree I did when our website was hacked the first time – all I had was a brochure that described our vanished archival project. Yet, magically it seems, the digital remains, much as I claim elsewhere that the repertoire does (Taylor, 2003), (almost) always there, though not always available to vision. The illusion of spontaneity and fleetingness we get from our email exchanges is only an illusion. The deleted email, it seems, can always come back to haunt us.

And what about embodiment? Embodiment as in 'having a body,' thinking, working, remembering, and expressing through our bodies, is central to performance studies, where we focus mainly (though certainly not exclusively) on incorporated behaviors and practices. Archives have inanimate bodies – the living author, as creator of a literary work, for example, becomes a literary corpus in the library. Online, of course, the blurring becomes more intense. Pure personas, unencumbered by our persons, have full play. 'On the Internet,' as one *New Yorker* cartoon reminds us, 'nobody knows you're a dog' (Steiner, 1993). These questions are not only the matter of theoretical discussion among scholars. Artists want to know how their performances will change when they are available online through streaming video. The first life of performance, the one they invest in, makes a clear artistic and/or political intervention. It finds its audience and creates a community of spectators/participants in the here and now. The second life of performance, for them, might be the edited, commercial version of work available for sale on DVD or video. This lacks the element of presence, but it is still marketed to specific audiences. Artists control the quality of the performance through editing and other processes. The third life of performance – the archival life of online streaming video – takes the long, historical view: 'Preserving performance for 500 years.' Who will watch these videos, and what will viewers make of them? As political performer, Reverend Billy, put it, 'You go forward and it [the video of the work] is still there. People take it into their lives in the present tense' (Reverend Billy (i.e., Bill Talen), 2006). Yet they, like scholars, understand the importance of archival materials – how can we understand Chicano theatre or popular theatre movements in the United States, without having access to the collections of El Teatro Campesino? Where will we turn to find original footage of work by performance artists? These collections include not just the glossy productions – in fact the HIDVL preserves only the original footage, not the edited or commercial products.

While working on the many lives of performance, I have come to believe we may be helping to shape academic fields both in the 'south,' where performance studies is gradually taking hold, and in the 'north,' where scholars are beginning to discuss 'Hemispheric' along-side American and Latin American studies. As Mexican theorist Rossana Reguillo argues, 'it has never been as vital as it is now to break the geographical determinants of thought' (Reguillo, 2005: 25; translation Taylor). The Americas share a complicated, back-and-forth history of migratory, cultural, economic, political, linguistic, and religious prac-tices. A hemispheric focus requires that we look at the Americas not only as a series of independent states, or as a geographical fact, but as the enacted and contested arena of these criss-crossings and encounters. A hemispheric focus would allow us to see Buenos Aires and New York City as major centers in the Jewish Americas, with places like 'Hotel Bolivia' (Spitzer, 1998) as stops along the routes of migration. How can we understand the formations and performance of race in the Americas without understanding the caste system imposed by the Inquisition in the sixteenth century that became codified in such different ways in the United States, Mexico, Peru, Brazil, Argentina, the Caribbean, to name a few? Performance traditions travel when communities leave home, either for economic or political reasons. Youth gangs span the Americas, transmitted by the Latino youth who grew up in the United States and deported for immigration reasons to a 'home' in Latin America they never knew. These histories and practices cannot be understood from any one place. They can certainly not be understood within the confines of the United States or American Studies, as expansive and encompassing as they might be.

What the Hemispheric Institute offers, then, is active engagement and dialogue (however complicated) among those who are seriously commit-ted to translocal, multidisciplinary work on performance and politics in the Americas. Our institutional and individual members have worked together for years now, creating something that was not there before – an online and offline environment of learning that brings scholars, artists, and activists throughout the Americas together to extend each other's work. That has been a wonderful performance practice. However, the effects of this practice will hopefully be felt long term. In 2007, the Hemispheric Institute received an endowment from the Ford Founda-tion and New York University to keep the project going well into the future. As or more important, however, all this collaborative work has found its way back into institutional homes throughout the Americas, where scholars are creating departments and programs that focus on performance research and practice. While performance practices – from art to activism – have always been a vital part of social and political life in the Americas, they are now increasingly becoming a vital part of academic life.

Notes

1. See Epstein, 1990, for a full description of this project. Osvaldo Dragún died in 1999.
2. For an excellent study of this history, see Grandin, 2006.

References

Escuela Internacional de Teatro de America Latina y el Caribe, 'Nuestra Utopia,' *EITALC website*, http://usuarios.lycos.es/eitalc/objetivos.htm (accessed 13 December 2008).

Epstein, S. 'Open Doors for the International Theatre School of Latin America and the Caribbean,' *The Drama Review (TDR)*, 34:3 (Autumn 1990) 162–76.

Grandin, G. *Empire's Workshop: Latin America, the United States, and the Rise of the New Imperialism* (New York: Metropolitan Books, 2006).

Hemispheric Institute of Performance and Politics, 'Mission,' *The Hemispheric Institute of Performance and Politics website*, http://www.hemisphericinstitute.org/eng/aboutus/index.html (accessed 13 April 2009).

McKenzie, J. *Perform or Else: From Discipline to Performance* (London and New York: Routledge, 2001).

Reguillo, R. *Horizontes fragmentados* (Tlaquepaque, Mex.: Instituto Tecnológico y de Estudios Superiores de Occidente, 2005).

Reverend Billy (i.e., Bill Talen). 'Performance and Politics in the Americas,' Guest lecture convened by Diana Taylor, Faculty Resource Network, New York University, 14 June 2006.

Spitzer, L. *Hotel Bolivia: The Culture of Memory in a Refuge from Nazism* (New York: Hill & Wang, 1998).

Steiner, P. 'On the Internet, Nobody Knows You're a Dog,' *The New Yorker*, 69:20 (5 July 1993) 61.

Taylor, D. 'Translating Performance,' *Profession 2002* (2002) 44–50.

———. *The Archive and the Repertoire: Performing Cultural Memory in the Americas* (Durham, NC: Duke University Press, 2003).

2
Interdisciplinary Field or Emerging Discipline?: Performance Studies at the University of Sydney

Gay McAuley

At the final session of the PSi (Performance Studies international) conference in Singapore in 2004, conference co-organizer Ray Langenbach reported that preliminary analysis of the participants indicated that 20 percent were locals living in Singapore, 40 percent came from other Asian countries, 15 percent from Europe, and 25 percent from America. Someone called out from the floor, 'Where did you put Australia?' The laughter provoked was intensified when Ray admitted that he did not know the answer. This incident reveals in an almost emblematic way the ambivalent position of Australia in relation to dominant geopolitical and cultural forces at the beginning of the twenty-first century. Is Australia part of Europe (the majority of its citizens are of European descent and its institutions are overwhelmingly British in origin), or part of the geographical region (its closest neighbors are Papua New Guinea and Indonesia)? Or has it now become a virtual outpost of the United States, as former Prime Minister John Howard implied when, to the dismay of most Australians, he expressed himself willing to be seen as George Bush's deputy sheriff in the region?[1] It was, of course, an Australian who asked the question, for it is Australians who experience most directly the uncertainties as to where their country is located – and is seen to be located – within the global polity, and where they themselves would locate it. The gap between the view or views from within and views from without indicates the complexity of the country's current colonial/postcolonial situation and the way this impacts upon relations with both geographical neighbors and erstwhile colonial authorities.

I begin with this anecdote because it resonates with my topic in this chapter concerning the institutional structures and conceptual frameworks that have shaped performance studies at the University of Sydney and the relations of similarity and difference that can be traced between the Sydney program and dominant traditions of performance and theatre studies in the United Kingdom and the United States, countries with which Australia has deep cultural ties. The (post)colonial relationship impacts on the way performance studies has been conceived and developed at Sydney, on the way it locates itself in relation to developments elsewhere, and on the way it is seen by others. Or, indeed, not seen. In most of the genealogies of performance studies that have been published, whether these emanate from US- or UK-based scholars, the department is not mentioned even though it has been in existence for nearly 20 years. It was, therefore, particularly gratifying to read Ian Maxwell's description of the Sydney department's history, published in the fiftieth birthday edition of *The Drama Review* (*TDR*) (Maxwell, 2006: 33–45), alongside articles recounting the interconnected development of the journal and the formation of performance studies as an academic discipline at New York University (NYU).

The editorial gloss put on the article was, however, slightly troubling. The descriptive note in the table of contents states that: 'Performance studies at the University of Sydney has developed in relative isolation from the "mainstream" of performance studies in both North America and Europe' (*Drama Review*, 2006: 2). This creates a center/periphery framework, prompting readers to locate Sydney at the periphery. My response to the statement is, first, to question what is meant by the 'mainstream' of performance studies in Europe (most European languages do not even have a word for performance and the dominant institutional location for the study of contemporary performance is departments of theatre studies); and, second, to point out the loaded nature of the term 'isolation.' If globalism is to be anything more than a euphemism for a renewed form of colonial domination, it surely involves recognition that in today's world there are many centers and many peripheries, and it requires a commitment to unpacking the ideological assumptions underpinning claims that a given place is isolated (from what center?). I hope that this chapter will indicate how intensively the Sydney-based scholars engaged with work being done elsewhere as they developed their own distinctive approach to the emerging discipline, the debt they owe to others, as well as the areas in which they can legitimately claim to be pioneers.

In the beginning

Performance studies was introduced at the University of Sydney in 1988 as an interdepartmental course (IDC) at the third-year undergraduate level, a fulltime honors' year was added in 1989, and postgraduate research students enrolled in significant numbers throughout the 1990s. The teaching program was preceded by at least ten years of experimental projects involving professional theatre practitioners having their work observed, analyzed, documented, and explored in a variety of teaching and research contexts. The administrative basis and funding for these projects came from the Theatre Workshop, a small unit that, under the inspired leadership of Derek Nicholson, facilitated collaboration between theatre artists and academics who were, for the most part, teaching theatre courses in language and literature departments. In the early years of the IDC, teaching was carried out by academics seconded for a fraction of their time from other departments in the Faculty of Arts: Tim Fitzpatrick from Italian, Terry Threadgold from English, Tony Day from Indonesian and Malayan Studies,[2] and me from French.

We were supported by an informal advisory group consisting of other colleagues from those four departments as well as others (notably Music, German, Architecture, Classics, and Anthropology) on whom we could draw for advice, occasional lectures, and participation in projects feeding into both teaching and research. The first direct appointment to performance studies was anthropologist Lowell Lewis, appointed in 1991 to a joint lectureship in Anthropology and Performance Studies, thus maintaining the principle of interdepartmental collaboration and filling what we saw as the major lack in the expertise we were bringing to bear on the phenomenon of performance. The Centre for Performance Studies was established in 1989, bringing together the new interdepartmental unit responsible for teaching and research in performance studies and the former Theatre Workshop.

Looking back to the introduction of performance studies and the developments that have occurred since, three crucial factors stand out. First was the experience of observing the creative process of professional theatre artists, made possible through the projects funded by the Theatre Workshop; second was the particular combination of scholarly interests and expertise brought together by the four academics, notably continental semiotics and reception theory, discourse analysis based on Michael Halliday's functional-systemic linguistics, theatre history, and feminist theory; and third was Tony Day's continual insistence that assumptions emanating from the history of European theatre be problematized by reference to performance practices located in other cultural centers, notably Indonesia (his own field of expertise) and Japan (with the assistance of Allan Marett from Music). These three factors, together with the theories of culture and fieldwork methodology introduced through our involvement with anthropology, are central to the practice of performance studies as it has developed at Sydney University over the last 20 years.

Shannon Jackson has commented, with great acuity, on 'the intense institutional work that those of us in performance studies have had to do in order to create a place for our work and our students' (Jackson, 2006: 29). The point needs to be emphasized because the institutional contexts within which performance studies has maneuvered to find its place are another significant factor that is shaping the discipline as it develops in different locations. Recent accounts of the emergence of performance studies programs in the United States and in the United Kingdom, their relationship with theatre and drama departments and the struggle of these for academic legitimacy, provide food for thought concerning the force of institutional structures and the impact of historical juncture on intellectual developments.[3]

Establishing performance studies at Sydney as an IDC was a pragmatic decision, as there was no likelihood of our being accorded permission (either from the federal government that funded the university or from the university governing body, or even from the Faculty of Arts) to establish a new department. From our perspective, however, this was a positive advantage in that it facilitated the interdisciplinary approach that we favored. The program had to be supported from within the existing staffing arrangements – hence the proposal that academics would be seconded for a fraction of their teaching load to the IDC. The IDC idea was itself a radical move at the very conservative University of Sydney, where traditional disciplines were enshrined in departments, led by titular professors whose responsibility to profess their discipline was sanctioned by government regulation. Before I could get permission to introduce performance studies, I had to establish the mechanism within the Faculty of Arts whereby such full 'professor-less' courses could be administered. This involved several years of preliminary committee work and advocacy, but the historical juncture was on our side as, in the 1980s, interdisciplinarity and challenges to disciplinary boundaries were recognized by all bar the most conservative to be the cutting edge of humanities research. However, as Marvin Carlson has warned, 'almost every aspect of university organization operated and still operates according to disciplinary lines – hirings and promotions of faculty, admission and training of students, representation on administrative committees' (Carlson, 2001: 139). The conservative force that such departmental structures exert on disciplinary and interdisciplinary development needs to be recognized, and departments such as performance studies have to be alert to the risk that, in return for the security of a (very small) designated staffing base, they may jeopardize the interdisciplinarity that structures such as the IDC made possible.

Another factor that has influenced the development of performance studies has been the huge range of technological advances that have occurred over the period. It is salutary to reflect that in the 1960s, when research into performance began seriously, we did not have video cameras, classrooms were not equipped with television monitors, and there were no photocopiers, let alone computers or internet facilities. Research, teaching, our ideas of what constitutes a document and our responsibility towards documentation, especially in performance research, have all been transformed by advances in technology. An interesting feature of this unremitting technological flow is that universities have been more willing to purchase equipment than to fund teaching positions. Performance studies at Sydney thus has been relatively well resourced in terms of equipment, permitting us to establish a substantial archive of performance documentation, while being seriously understaffed throughout its existence. It has also become evident that recording and photographing performance involve the acquisition of new skills, both in relation to the making of useful records and to retrieving the information they hold concerning the live event. Training in these skills is now an integral part of the performance studies program.

The institutional and disciplinary context

When performance studies was introduced at the University of Sydney, there were no departments or programs in performance studies at any other Australian universities, and to our knowledge, the only other full department in existence was at NYU. Departments of drama and theatre studies were themselves relatively new in Australia, having been set up in several universities in the 1960s and 1970s, and these, with the exception of Flinders University in Adelaide, followed the model established in Britain, notably by the University of Bristol. Central to the research and teaching program in these departments was theatre history, dominated by what Marvin Carlson has called 'the traditional, European canonical works of high literary art' (Carlson, 2001: 138), approached through a combination of historiographical methods and practical, workshop-oriented explorations, usually based on dramatic texts. Apart from Flinders, which began to offer actor training a few years after its establishment in 1967, drama departments did not then undertake practical training for work in the theatre. This was provided by conservatories and specialized institutions, in particular the National Institute of Dramatic Art (NIDA), which was located at the University of New South Wales, on the same campus as the country's first drama department, even sharing some resources with that department while maintaining a rigid demarcation between the two (see Jordan, 1995: 619–21).

While my colleagues and I had no ambition or desire to compete with NIDA in the provision of vocational training, we did have reservations about the adequacy of the approaches to theatre and performance offered by the dominant theatre and drama studies model. Performance studies, as we conceived it, was intended in the first instance to provide an alternative approach to this model, which seemed to us to be too narrowly focused on the history of European theatre, particularly that of the English-speaking diaspora, and even within that field tended to pay too little attention to performance traditions that had not left textual residues of high literary merit. Furthermore, NIDA did not seem to be open to engaging seriously with the stream of performance work that had been emerging throughout the 1970s and 1980s in Sydney, as in many other places, alongside traditional text-based theatre practice. Just as problematic from our point of view was that colleagues in theatre studies departments were following the lead of their British counterparts and were uninterested in, or actively hostile to, the new structuralist, semiotic, and poststructuralist theories that had revolutionized the study of communications of all sorts in continental Europe. These seemed to offer new languages with which to describe and analyze performance, as well as means of deepening our understanding of the relationship between society, text, and performance. This was the institutional and disciplinary context within which performance studies at Sydney sought to position itself. Our goal was to study a wide range of performance practices, including social and cultural performances, ritual, sport, and other activities with a performative dimension, as well as genres of aesthetic performance, to bring some new critical and analytical methodologies to bear on this material, and to base our theoretical reflections on detailed observation of the creative processes of theatre and performance artists.

Insisting on engagement with contemporary performance practice via performance analysis and observation of the rehearsal process is probably the single most important area of difference between the Sydney version of performance studies and approaches adopted in other major centers. It raises thorny issues concerning the nature and function of performance practice in performance studies as well as in theatre studies. Whose practices are being researched? How carefully is evidence collected and stored? To what extent are findings replicable? Are people developing empirically based research or are they engaged in purely interpretive studies? One of the major innovations of theatre studies pedagogy has been its reliance on workshops and performance practice, undertaken with students, in order to demonstrate how theatrical meanings are made, the part played by the text within this process, and so on. This very valuable pedagogical practice is not without its dangers, for it can function to lock students into a narcissistic preoccupation with their own creative processes to the exclusion of any engagement with performance culture outside the sheltered walls of the university. Furthermore, scholarly research based on this sort of in-house performance practice rarely acknowledges that there is a gulf between the kinds of performance produced by untrained students, however eager and gifted, and the depths of insight reached by experienced professional theatre artists. This is why performance studies at the University of Sydney is based exclusively on collaboration with practicing artists: students do not make performance works but observe, analyze, document, and theorize about the performance practices of professional artists. Some postgraduate students are themselves performance practitioners whose research may draw on their own practice, and these accounts from 'within' are valuable, especially in an art form in which practitioners have, until recently, rarely written in any depth about their work. However, as contemporary ethnographers have taught us, insider accounts need to be complemented by outsider accounts, and to do justice to the complexities of the production/performance/reception process requires multiple perspectives and multiple theoretical frameworks.

Richard Schechner's advocacy for the NYU department he founded was extremely useful to us as we sought academic respectability for the program we were proposing. We pointed to Schechner's writings, particularly his editorials in *TDR*, in which he put the case for a performance studies that could deal with a 'broad spectrum' of performance ranging from ritual to everyday behaviors, and that would locate varied aesthetic performances within this web of activities (Schechner, 1985, 1988). Arguing that performance was the coming thing, we referred to Schechner to support the need to include a far wider range of performance activities than had been studied in departments of drama and literature, and stressed that deep understanding of performance processes would be required if performance was to be seen as a theoretical category through which other forms of social activity could be analyzed. In a short chapter, it is not possible to acknowledge all the influences that shaped our early thinking, but alongside Schechner mention must be made of theatre semioticians such as Anne Ubersfeld (1977, 1981), Patrice Pavis (1980), and Marco de Marinis (1982, 1985), as well as the work of sociologists and anthropologists Erving Goffman (1974), Victor Turner (1982, 1987), and Clifford Geertz (1973, 1983). Terry Threadgold's *Feminist Poetics* was not published until 1997, but her ideas were developing over the period in question, as indeed was my own appreciation of the importance of spatial factors in the performance experience (McAuley, 1999). Locating the performance event within its cultural and social contexts, exploring the role of the various agents responsible for making the performance, and taking seriously 'the aftermath' of the performance (Schechner, 1985) are central to the task of performance studies. The shift of emphasis from regarding performance as a product to a series of processes was encapsulated in the title Tim Fitzpatrick gave to the collection of essays he edited (Fitzpatrick, 1989), emerging from the research we had been doing in the years immediately preceding the establishment of the IDC.

A key difference between performance studies as it evolved at Sydney University and the NYU model was that we never developed what Richard Schechner himself later came to see as the 'weird blind spot' with regard to mainstream theatre (Schechner, 2000: 4–6). While the interests of academic staff and postgraduate students at Sydney have ranged widely over performance genres as disparate as postmodern performance, contemporary hip hop, *commedia dell'arte*, *wayang kulit*, restorative justice practices in the juvenile justice system, and opening ceremonies of the Olympic Games, a good deal of attention has always also been paid to processes and practices involved in the creation and reception of traditional text/narrative/character-based theatre. Acknowledging the centrality of this tradition in the culture of the West does not entail ignoring other traditions of high art, or popular performance traditions, or the practices of artists in the West who have contested the mainstream, and it certainly does not mean that social and cultural performances need to be regarded as peripheral. Two factors have limited the extent of coverage of performance genres and practices in our teaching and in the research we have supervised: first, the small number of academic staff;[4] and, second, our awareness that substantial cultural and linguistic competence are required for study of any performance practice. Here the experience of those trained in anthropology joins with those whose primary background was in foreign language departments to view with great suspicion assertions about performance that are not grounded in deep cultural knowledge, or assumptions of cultural knowledge that are not based on competence in the relevant languages.

There is a tendency in performance studies to cast the net wider and wider, accepting an ever-expanding range of performance practices as legitimate objects of study. While such openness has its attractions, there are problems with the notion of a 'field without limits'; it seems to me that even though understandings of what constitutes performance may differ from culture to culture and over time, we do need to define with some care what we mean by it here and now. My own rule of thumb has been that for an activity to be regarded as performance, it must involve the live presence of the performers and those witnessing it, that there must be some intentionality on the part of performer or witness or both, and that these conditions in turn necessitate analysis of the place and temporality which enable both parties to be present to each other, as well as what can be described as the performance contract between them, whether explicit or implicit.[5] The major task of performance studies, as I see it, is less to preside over an endlessly expanding range of objects of study, but rather to develop the practical and methodological bases for serious study of whatever practices are currently conceptualized as performance, and to make clear what it is that a performance studies approach brings to bear on its objects of study that, say, a cultural studies approach or even a theatre studies approach does not. If we cannot do this, then our claims to disciplinary status are severely weakened.

Coming to performance studies with a longstanding interest in the theatre, my personal aim for the 15 years during which I led the department was to explore and refine a range of methodologies that would enable us to engage at a serious analytical level with the whole production/performance/reception process in a wide variety of aesthetic performance practices. The task involved in the first instance was finding and fostering a network of artists who would collaborate with us, permit us to observe, document, and analyze their creative processes and the performances that resulted, as well as enter wholeheartedly into the attempt to explore and test hypotheses about making and communicating meaning through performance. It also involved developing methods of documentation and devising formats that would enable such work to be brought into the rigid framework of a university timetable. Observing and documenting the rehearsal process, noting in particular the issues that were most discussed and the terms used by the practitioners, and documenting and analyzing the resulting performances revealed much about the ways in which meaning is made and communicated in the theatre and about the creative agency of those involved. It also revealed, as Terry Threadgold has argued, a great deal about identity construction and the presentation of self in contexts far beyond the rehearsal room or the stage (Threadgold, 1995: 172–82).

It is perhaps evident from this brief account that the conceptual framework underpinning the investigation was dominated by semiotic theory, and it is true that at the beginning my interest in rehearsal functioned largely as a means to gain a deeper and more complex understanding of the resulting performance. As time went on, however, I began to realize that rehearsal needed to be studied on its own terms and not simply as a means to deeper appreciation of the end product. This has led to the development of rehearsal studies, which has entailed a shift from projects where artists explored questions posed by the researchers to finding opportunities for researchers to become participant observers in companies making performance, and to observe the full rehearsal process leading to performance. It has also entailed broadening the methodological and conceptual framework for studying performance through the application and adaptation of approaches and concepts borrowed from ethnography. The value of the work has been recognized to the extent that the observation and analysis of rehearsal process is now a foundational element in the research training provided by the Sydney department.[6] Ethnography has been used in this case not, as is frequently the case in performance studies, to explore performance practices in cultures far removed from the home base, but to find ways to deal with the multiplicity of culturally embedded processes involved in a practice that is central to the home culture and for which the dominant theatre studies approaches provided little guidance.

Looking to the future

There has certainly been a marked increase in the number of performance studies programs being offered in tertiary institutions in the English-speaking world, although the number of full departments is still relatively small. The peak professional body in the field in Australia, formerly the Australasian Drama Studies Association, now proclaims on its letterhead that it is the Australasian Association for Theatre, Drama and Performance Studies, although it still retains the acronym ADSA. This development is characteristic of the way performance studies has spread: departments of drama or theatre studies add the word 'performance' to their title, scholarly journals include it as a subtitle, and theatre studies conferences have introduced focus groups and panels dealing with performance and performance studies. In many instances these name changes simply indicate institutional recognition for a wider range of aesthetic performance practices than had hitherto been the case – but even this expansion of the field is a significant achievement. It is an acknowledgment that the many modes of aesthetic performance included under the umbrella category 'performance' are in fact an integral part of contemporary theatre practice rather than completely separate cultural phenomena. With hindsight, it is becoming clear that the performance genres developed over the last 30 years represent another twist in the age-old story of theatre resisting predictions of its imminent demise and being reinvented to reveal in new ways the potency of live performance.

It is less clear whether the name changes indicate any firm consensus about the inclusion of social and cultural performance more broadly – whether, for example, performative behaviors in daily social interactions or in sport, worship, litigation, or war come within the purview of all these newly rebadged departments. It does seem, however, from the spate of recently published introductory course books, collections of canonical texts, and translations into English of seminal works by European theorists that there is some agreement emerging concerning the range of explanatory theories and analytical methodologies appropriate to deal with the extended field. Ongoing debates about the disciplinary status of performance studies certainly reflect uncertainties and tensions in its relations with other scholarly approaches to similar fields of human activities. In this respect, performance studies is no different from other disciplines in the humanities and social sciences, where the impact of postmodernism has provoked deep internal debates about epistemology, genealogy, and increasingly blurred disciplinary boundaries. Proponents of performance studies have always insisted on its interdisciplinarity, but increasingly the idea of 'between-ness' has been replaced by a more radical sense that the notion of the discipline as an organizing structure of knowledge has broken down and, for some, performance studies is a prime example of the new 'postdisciplinarity' (Case, 2001: 145–52). However, as Marvin Carlson has reminded us, universities are organized along disciplinary lines, with disciplines represented by departments, and it will not be easy for post-disciplines to survive, never mind actually grow, outside the disciplinary/departmental structure. The success of performance studies, measured by the extent to which it has achieved institutional recognition, brings with it new responsibilities: the attractions of post-disciplinarity notwithstanding, offering an undergraduate major or postgraduate research training in performance studies is a tacit acknowledgement that it is indeed a discipline, albeit an emerging one, and we therefore have to face up to the new demands that this creates.

Reflecting on the work being undertaken by the second generation of performance studies scholars at the University of Sydney, most of them trained in performance studies rather than in one of the contributing disciplines, I am optimistic about the future although aware of the challenges that they face in a field dominated by northern hemisphere institutions and perspectives, reinforced by a market economy within which the Australian experience is so often seen as utterly marginal. Current research in the department includes Paul Dwyer's work on social memory and his project on restorative justice in Bougainville, Ian Maxwell's use of phenomenology to investigate sport spectatorship and theories of acting, Amanda Card's exploration of hybridity and appropriation in dance practices in colonized societies, and Laura Ginters's participation in a research cluster dealing with rehearsal practices in theatres of the past 400 years. In this work and in the wide range of doctoral research they supervise, my colleagues are extending and refining the methodologies developed over the last 20 years and making their own contribution to an increasingly sophisticated critical engagement with performative practices throughout society. The stakes are high for them and others around the world as they strive to retain the openness that comes from operating at the interface of many practices and many discourses while achieving institutional recognition within a disciplinary structure.

Notes

1. Much has been written about this controversial comment, but it appears to have originated in an interview with John Howard by journalist Fred Brenchley published in *The Bulletin*, 29 September 1999.
2. Now renamed the Department of Indonesian Studies.
3. See, for example, Fischer-Lichte, 1997 and 1999; Jackson, 2004; McKenzie, 2001; Shepherd and Wallis, 2004; and Shevtsova, 2001.
4. Adding to the problems intrinsic to launching an interdisciplinary venture, performance studies at Sydney University has had to contend with a period of drastic reduction in government funding to the universities which has seen the number of full-time academic staff in the Faculty of Arts greatly reduced over the last 15 years. As the traditional disciplines have struggled to maintain a reasonable coverage of their core subject matter, it has become increasingly difficult to obtain new positions to develop interdisciplinary fields.
5. For a more detailed discussion of the concept of 'liveness,' its centrality in my understanding of performance and a response to Philip Auslander's argument concerning the relation of live and mediated, see McAuley, 2001: 5–19.
6. Some results of this work can be seen in the 2006 issue of *About Performance* (special issue on 'Rehearsal and Performance Making Processes'), published by the Department of Performance Studies, University of Sydney – see McAuley, 2006. Also see also McAuley, 1998: 75–85; and McAuley, 2008.

References

Carlson, M. 'Theatre and Performance at a Time of Shifting Disciplines,' *Theatre Research International*, 26:2 (2001) 137–44.

Case, S.-E. 'Feminism and Performance: A Postdisciplinary Couple,' *Theatre Research International*, 26:2 (2001) 145–52.

De Marinis, M. *Semiotica del Teatro: l'analisi testuale dello spettacolo* (Bologna: Bompiani, 1982).

———. *L'esperienza dello spettatore: fondamenti per une semiotica delle ricezione teatrale* (Urbino: Centro Internazionale di Semiotica e Linguistica, 1985).

Drama Review, The. Table of Contents, *The Drama Review (TDR)*, 50:1 (2006) 2–3.

Fischer-Lichte, E. *The Show and the Gaze of Theatre: A European Perspective*, ed. and trans. Jo Riley (Iowa City: University of Iowa Press, 1997).

———. 'From Text to Performance: The Rise of Theatre Studies as an Academic Discipline in Germany,' *Theatre Research International*, 24:2 (1999) 168–78.

Fitzpatrick, T. (ed.) *Performance: From Product to Process* (Sydney: Frederick May Foundation for Italian Studies, 1989).

Geertz, C. *The Interpretation of Cultures* (New York: Basic Books, 1973).

———. *Local Knowledge: Further Essays in Interpretive Anthropology* (New York: Basic Books, 1983).

Goffman, E. *Frame Analysis: An Essay on the Organization of Experience* (Boston, MA: Northeastern University Press, 1974).

Jackson, S. *Professing Performance: Theatre in the Academy from Philology to Performativity* (Cambridge: Cambridge University Press, 2004).

———. 'Caravans Continued: In Memory of Dwight Conquergood,' *The Drama Review (TDR)*, 50:1 (2006) 28–32.

Jordan, R. 'Universities,' *Companion to the Theatre in Australia*, ed. P. Parsons (Sydney: Currency Press, 1995), pp. 619–21.

Maxwell, I. 'Parallel Evolution: Performance Studies at the University of Sydney,' *The Drama Review (TDR)*, 50:1 (2006) 33–45.

McAuley, G. 'Towards an Ethnography of Rehearsal,' *New Theatre Quarterly*, 14:1 (1998) 75–85.

———. *Space in Performance: Making Meaning in the Theatre* (Ann Arbor: University of Michigan Press, 1999).

———. 'Performance Studies: Definitions, Methodologies, Future Directions,' *Australasian Drama Studies*, 39, special issue on 'Performance Studies in Australia' (2001) 5–19.

———. (ed.) *About Performance*, 6, special issue on 'Rehearsal and Performance Making Processes' (2006).

———. 'Not Magic but Work: Rehearsal and the Production of Meaning,' *Theatre Research International*, 33: 3, special issue on 'Genetics of Performance,' guest ed. J. Féral (2008) 276–88.

McKenzie, J. 2001. *Perform or Else: From Discipline to Performance* (London and New York: Routledge, 2001).

Pavis, P. *Dictionnaire du Théâtre* (Paris: Messidor/Editions Sociales, 1980).

———. 'Theatre Studies and Interdisciplinarity,' *Theatre Research International*, 26:2 (2001) 153–63.

Schechner, R. *Between Theatre and Anthropology* (Philadelphia: University of Pennsylvania Press, 1985).

————. 'Performance Studies: The Broad Spectrum Approach,' *The Drama Review (TDR)*, 32:3 (1988) 4–6.

————. 'Mainstream Theatre and Performance Studies,' *The Drama Review (TDR)*, 44:2 (2000) 4–6.

————. *Performance Studies: An Introduction* (London and New York: Routledge, 2002).

Shepherd, S. and M. Wallis. *Drama/Theatre/Performance* (London and New York: Routledge, 2004).

Shevtsova, M. (ed.) *Theatre Research International*, 26:2, special issue on 'Theatre and Interdisciplinarity' (2001).

Threadgold, T. 'Postmodernism and the Politics of Culture: Chekhov's *Three Sisters* in Rehearsal and Performance,' *Southern Review*, 28:2 (1995) 172–82.

————. *Feminist Poetics: Poiesis, Performance, Histories* (London and New York: Routledge, 1997).

Turner, V. *From Ritual to Theatre* (New York: PAJ Publications, 1982).

————. *The Anthropology of Performance* (New York: PAJ Publications, 1987).

Ubersfeld, A. *Lire le Théâtre* (Paris: Editions Sociales, 1977).

————. *L'Ecole du Spectateur* (Paris: Editions Sociales, 1981).

3
The Practice Turn: Performance and the British Academy

Heike Roms

Preliminary: performance anxiety

It is the spring of 2007, and although this chapter has not yet been written I already know it will have been published too late. All university-based research currently undertaken in the United Kingdom, in whichever discipline, is done so in anticipation of the forthcoming *Research Assessment Exercise 2008* (also known as 'RAE 2008'), a national audit of research activity, conducted by the UK's Higher Education funding councils, that will for the next few years determine the level of research funding that the councils allocate to each British university.[1] Through a process of expert review, institutions will be ranked principally on the quality of the 'output' of their research-active staff. This chapter was to be one of my outputs, but the inevitable delays that occur in a scholarly collaboration across continents have now pushed the prospective publication date of this book beyond the census date of the audit. Other scholars across Britain are currently still waiting anxiously to see whether their book or chapter or journal article will be printed in time as publishers work through a backlog of manuscripts.

For performance scholars, the *Research Assessment Exercise* presents a rich subject into which to enquire. First introduced in 1986 under a Conservative government, the exercise, now conducted for the sixth time,[2] suits the performance-oriented approach to teaching and research that the present New Labour administration has been promoting across the educational sector.[3] Work is judged 'in terms of originality, significance and rigour' (HEFCE et al., 2005: 31) against a benchmark of 'world-leading' 'excellence' (ibid: 5).[4] In *The University in Ruins*, Bill Readings (1996) offers a pointed analysis of the discourse of 'excellence' that defines our contemporary model of the university. It indicates the university's transformation from an institution whose primary role was to promote the idea of a national culture into a transnational corporation driven by techno-bureaucratic interests. In this discourse, qualitative judgments often conceal quantitative concerns. Indeed, the importance of considerations such as return on capital (in a global scholarly market) for the *raison d'être* of the *Research Assessment Exercise* is clearly expressed in its publications: the summary report of the last RAE in 2001 boasts proudly – although ambiguously, implying as it does that research in Britain is under-funded – that 'UK researchers are among the most productive, and the number of times their work is read and used by other academics per million pounds spent is the highest in the world' (HEFCE et al., 2001: 2).[5] Reader, you may cite me on this.

Only months away from the audit date a certain performance anxiety is growing among the British academic community: the most frequently expressed concerns are pressures on productivity, worries about the consequences for one's career, and unease about the implications of the performance indicators – which differentiate between 'world-leading,' 'international,' and 'national' – for scholars who are invested in localized research (although the RAE has assured that the indicators refer to 'quality standards' and not to the 'geographical scope' of research; HEFCE et al., 2005: 31). But whatever one's misgivings about the RAE, there are also gains to be had from its current format. An evaluation based primarily on the academic judgment made by a panel of one's peers,[6] elected from among the scholarly community, with an emphasis on the quality of selected items rather than on the overall volume of research production, is designed not to prejudice submissions from scholars and departments that are not part of a traditionally research-led elite institution. For the government, the efficacy of this labor- and cost-intensive process has itself now been called into question, and there are plans to replace it with the so-called *Research Excellence Framework*, a system in which qualitative judgment is to be (at least to a large part) replaced by quantifiable 'metrics,' measuring, for example, the 'impact' of published papers by the number of citations they engender.

Shifting one's focus from the performance of status to the status of performance in the *Research Assessment Exercise* reveals another feature of the current system which is of particular importance to the field of performance research: namely a relatively nuanced definition of what constitutes 'research' and how it may be evidenced. Since 1992,[7] for the purpose of the first full national RAE, research is characterized as including 'the invention and generation of ideas, images, *performances* and artefacts [...], where these lead to new or substantially improved insights' (UFC, 1992: 6; my emphasis). And recent guidelines on acceptable forms of dissemination list 'all forms of publicly available assessable output [...] likely to include [...] *performances*, compositions, designs, artefacts and exhibitions' (HEFCE et al., 2005: 20; my emphasis). This seemingly simple catalogue reveals that in the context of UK-based scholarship, the predominantly reception-based approach that characterized arts and humanities research in the modern university has been replaced by a model which regards creative-aesthetic practices, too, as forms of knowledge-making and knowledge-distribution that possess validity in academic terms.[8] This recognition is now widely institutionally embedded: it directs not only the operation of the RAE, but has been incorporated into the funding guidelines for the *Arts and Humanities Research Council (AHRC)*, the main funding body for arts and humanities research in the United Kingdom, and has led to the establishment of new protocols for practice-based (or practice-led or practice-as-research)[9] doctoral work.[10]

In the following I will briefly outline what I believe to be at stake in the recognition of the knowledge-making capacity of creative practice for the researching and teaching of performance in the United Kingdom. But, first, I will situate this debate within a wider discussion that underlies this chapter as it does this entire collection – namely how such localized concerns relate to the (American-born but increasingly globally active) discipline of performance studies.[11]

Performance research/performance studies: field or discipline?

It may be helpful at this point to draw a distinction between performance research as a field and performance studies as a discipline, if only as a working definition for the remainder of this chapter. Performance research as it is practiced in Britain today can, with some justification, be considered a field, albeit one which shifts between a loose community of individuals with differing research interests on the one hand, and a complex set of institutional structures that seek to define what is regarded as research in performance on the other. Performance studies, in contrast, has a strong claim to disciplinary status, with a range of associated terms, texts, and approaches, despite its own frequent assurances to the contrary. This claim has been made principally through the work of US-based scholars, above all Richard Schechner, who in his *Introduction to Performance Studies* attempts to provide what he calls

a 'wide-open field' (Schechner, 2002: 1) nonetheless with a genealogy, a set of methodologies, and a canon of significant literature. In
the United Kingdom, the process of disciplinary emergence has been
more diffuse. 'Performance studies' consequently has confusingly different current manifestations: it appears variously in the titles of British
university schools and departments (often preceded by 'theatre and'),
academic jobs, degree programs, and individual 'modules' (i.e., classes),
even as a school subject, and what is meant by it can range from a
synonym for 'performing arts,' to the study of interpretative and technical skills (particularly in the area of music practice), to the teaching
of 'devised' (i.e., collaboratively conceived) performance genres or of
performance art, to a research enquiry that shapes itself around the
debates we have come to associate with the discipline of 'performance
studies.'

It is the latter that I wish to focus on here. There have been several
recent attempts to map the development and distinctiveness of performance studies as an emerging discipline in the United Kingdom. In
the introduction to their *Routledge Companion to Theatre and Performance*
(2006), Paul Allain and Jen Harvie summarize its history as a continuation of familiar genealogies, namely the birth of performance studies
in the United States in its twin sites of New York University and Northwestern University (see elsewhere in this volume). The authors recount
this birth as a reaction to theatre studies' narrow focus on an artistic practice with limited cultural and social reach, and follow Richard
Schechner's proposition that the 'broad spectrum approach' (Schechner,
1988) of performance studies instead presents a 'paradigm shift' (Allain
and Harvie, 2006: 8; see also Schechner, 1992) towards greater interdisciplinarity and a new set of critical concerns. This shift has also taken
hold in the United Kingdom, where it allows scholars to widen their
approaches in addressing the problematic of public space, imperial legacies, or civic rights, for example. Even though, according to Allain and
Harvie, '[i]n Britain, performance studies appears to be in its infancy
in relation to the large family of drama and theatre departments that
exist,' they go on to suggest that 'whatever the titles of the courses on
which we teach, there is no denying the substantial impact this shift
from theatre to performance – to put it at its most crude – has had'
(ibid.).

This implies that in the United Kingdom, many drama and theatre
studies departments 'do' performance studies in all but name. Indeed,
Allain and Harvie speak of an 'increasing dominance of performance' and
a 'deepening entrenchment of performance and performance studies

in Britain' (ibid.: 9). So deep do they regard this entrenchment to be that the authors feel compelled to reinstate 'theatre' as a focus for their publication, in order 'to locate the book within a practice which is partly our history and in which performance studies' development is deeply embedded' (ibid.). Therefore, rather than replacing theatre studies altogether, as Schechner (1992) had predicted, the development of performance studies in the United Kingdom appears to have had the opposite effect, reinvigorating theatre studies through widening its objects and approaches of study, and thereby ultimately re-establishing 'theatre' as a theoretical, aesthetic, and political concern.[12]

In their editorial for a special issue on 'The practice of Performance Studies in the United Kingdom' for the journal *Studies in Theatre and Performance*, Jane Bacon and Franc Chamberlain (2005), too, credit Schechner's 'broad spectrum' model with having inspired a development that leads them to the observation that British drama and theatre studies departments are in the process of 'metamorphosing into Performance Studies departments' (ibid.: 185). But the authors emphasize that the growth of performance studies in the United Kingdom has taken place in very different circumstances from those in the United States: 'One of the key factors to be considered in the United Kingdom is the relationship of Performance Studies to Performing Arts and the problems that are faced by courses that attempt an interdisciplinary approach [...]' (ibid.: 181). The term 'Performing Arts' is here deliberately chosen – in his own contribution to the journal, Chamberlain accuses performance studies of prolonging the hegemony of a ' "theatre/ drama" model' (Chamberlain, 2005: 263) at the expense of the other performance arts such as music and dance. Interdisciplinarity in Bacon and Chamberlain's understanding is thus defined as a coming together of different artistic disciplines, rather than as the cutting across to the social sciences that Schechner had in mind. It is here where the two authors identify a potentially unique 'British' approach: 'We have a sense that something new is happening in the United Kingdom and that it does not yet have a name: it knows it is neither the Performing Arts collectively nor any of them individually, and that it is not the same as Performance Studies in the United States' (Bacon and Chamberlain, 2005: 187).

There are thus two features that seem to mark out performance studies in the United Kingdom: firstly, its attention to artistic practice, whether manifest as a renewed focus on the materiality of theatre or as the expansion of creative modes of investigation; and secondly, the tension between its close association with and frequent deliberate dissociation from its US-counterpart.

A special relationship?

In her 'Foreword from "across the pond" ' to Kelleher and Ridout's collection on *Contemporary Theatres in Europe* (2006) (a collection that features a majority of UK-based writers), American scholar Janelle Reinelt offers an assessment of the British scholarly scene from the perspective of her ' "Yankee" eye' (Reinelt, 2006: xiv) that reaches a similar conclusion to that of Allain/Harvie and Bacon/Chamberlain: she too proposes that 'the UK seems to be developing an understanding of performance studies that is related to American versions, but also unique to the British situation' (ibid.: xv–xvi). For her, this uniqueness can be found in a number of features: a 'strong community-based theatre tradition, born of deep commitments to place and heritage' (ibid.: xvi); the concept of 'performance as research' (ibid.); and the 'deployment of and involvement with [especially continental] philosophy' in place of the 'kind of identity politics and overt political activism familiar to American and Canadian theatre scholars' (ibid.: xvii).

Reinelt has subsequently moved across to this side of the pond[13] as part of a growing number of American scholars in theatre and performance studies who are reversing the traditional 'brain drain' that in the past attracted British academics to the United States with the prospect of higher payment and less administration. Nowadays, it is often their US-colleagues who are drawn to the United Kingdom by the increasing global visibility of British scholarship in the field, and by an expansion of theatre and performance departments and concomitant teaching positions resulting from a decade of consistently rising student numbers in the United Kingdom. Underlying such mobility in both directions has been the traditionally close relationship between US and UK scholarship, assisted by a shared language, long-standing cultural, economic, and political ties, and a compatible educational system. Indeed, in the case of performance studies it is important to note that Britain has often provided locations where what we now identify as the US-American foundational scholarship in the field was not just received, but actually produced. UK-based institutions such as the publisher Routledge[14] and the Centre for Performance Research have played an important role in the formation of performance studies in the United States and elsewhere. By the mid-1990s, Routledge had acknowledged the importance of this new area of research by changing the title of its list to 'Theatre and Performance Studies,' and it began publishing what were to become key texts for the formation of the discipline – and of course widely read also in the United Kingdom – by American authors such as Auslander, Diamond,

McKenzie, Phelan, Schechner, and Schneider. Alongside them, Routledge's catalogue has included significant works by British scholars that explore subjects of central importance to performance studies: questions of gender (Aston), site (Kaye; Pearson and Shanks), political efficacy (Kershaw), the everyday (Read), and others. (Tellingly, however, none of these have found their way into Schechner's performance studies textbook, which suggests that British-American scholarly exchanges are possibly not as established as they may appear when compared to the near-invisible status of non-Anglo scholarship in the field.)

An important conduit for such exchanges has been the Centre for Performance Research (CPR) (Christie, Gough, and Watt, 2006). The CPR transformed itself in 1995 from an independent theatre company founded in the early 1970s, to a university-hosted research organization[15] whose main contribution to the field has been its conferences, publication ventures, and performance library. From early on in its history, the CPR's intercultural approach to theatrical performance, inspired by its close affiliation with Eugenio Barba and the Third Theatre network, brought it into contact with the emerging discipline of performance studies, which it not only helped to promote in Britain through its *Points of Contact* conference series in the 1990s,[16] but, according to Schechner (2002: 16), also helped to define. In 1997, CPR's Artistic Director, Richard Gough, became the inaugural president of the new association for performance studies, *PSi Performance Studies international* (with myself as its first administrator),[17] and he went on to chair the fifth Performance Studies Conference in Aberystwyth in 1999, the first time the conference series took place outside of the United States.[18] Although thus instrumental in the formation of the discipline, Gough has of late distanced himself from what he characterizes as performance studies' 'dangerous and fecund' (Gough, in Christie, Gough, and Watt, 2006: 63) blurring of performance as episteme and performance as aesthetic practice. Instead, he reasserts CPR's position as a 'theatre organization' 'to the point where I would begin to question whether we do Performance Studies' (ibid.).

This ongoing tension between a close correspondence with and frequent deliberate dissociation from its US-counterpart mirrors and refracts the complexities of the wider 'special relationship' that exists between Britain and the States. I do not wish, however, to overlook Britain's position within Europe and its scholarly traditions, theatrical practices, and political preoccupations. What Reinelt has identified as a 'taken-for-granted seriousness of theatre as a cultural practice' (Reinelt, 2006: xiv), a continuing belief in 'the worth of the live event' (ibid.: xv), and a 'preoc-cupation with the sense of tradition and its reworking' (ibid.) in British performance scholarship can arguably be traced back to such European influences. Institutionally, too, Britain is part of the so-called *Bologna Process*, the European Commission's program to make the hitherto highly differentiated Higher Education systems in Europe more compatible by 2010.[19] As a consequence, the United Kingdom is increasingly attracting students and faculty (including myself) from other European countries, markedly in the area of theatre and performance where Britain has a greater number of courses and academic jobs on offer than any other European country. But as *Performance Research*, one of the leading per-formance journals edited and published in Britain,[20] with its keen sense of the locality of discourses and practices, proposed in its inaugural issue in 1995: 'Britain often feels as if it perches on the edge of Europe, some-where off the shores of America' (MacDonald and Allsopp, 1996: vi). It is this deeply implicated, and yet somewhat ex-centric, position in relation to both US and European scholarly discourses and institutional practices that has left its mark on how the study of performance is conducted and taught in the United Kingdom today.

Continued: performance anxiety

In *Drama/ Theatre/Performance*, Simon Shepherd and Mick Wallis (2004) offer a useful examination of the shifting terminology in the field by out-lining when and why each of the three eponymous terms gained critical currency within the context of the British university system, and identi-fying – not surprisingly – a 'successive emergence of Drama, Theatre and Performance as academic paradigms' (ibid.: 2). In summary, 'drama,' with its roots in the Western literary tradition, was the earliest concept to attain institutional recognition, which led to the (comparatively late) opening of the first university drama department in the United Kingdom in 1947 in Bristol,[21] whilst 'performance' has only latterly been recog-nized as a critical term around which both teaching and research are organized. But one concern has remained constant throughout the shift from drama to theatre to performance in the United Kingdom, namely what Shepherd and Wallis describe as persistent 'anxieties' (ibid.: 9) dis-played by the field around the proper role of creative practice within it. The authors trace these anxieties back to a 'nervousness, in the con-text of text-based university humanities departments, about both craft and vocational training' (ibid.). In contrast to the incorporation of voca-tional training in American universities and the continuing (if slowly dissolving) separation between the practice of art and its study in the European tradition, the British model therefore, with its attempted integration and mutual interrogation of practice and theory within research and pedagogy, may have created a situation of persistent anxiety, but it has arguably also produced an ongoing problematization of the position of performance practice within the academy that has proven highly productive.

Over the last 15 years or so,[22] this problematization has been concentrated around the notion of practice-as-research. It has been influenced significantly by the growing integration of professional artists into the university, furthered by the introduction of fellowships, a campus-based network of presenting venues, and, above all, the offer of paid employment in a time of unreliable public arts subsidies. The integration has been made easier by the fact that artists involved in innovative, interdisciplinary aesthetic practices such as live and performance art are often already engaged in research-based approaches, which distinguishes their work from the more product-focused activities of conventional theatre practice. (Indeed, so successful has the integration of performance artists into the academy become in the United Kingdom that a university-independent critical discourse around performance has not managed to take hold in the same way as it has in many other European countries, where venues or artists themselves often provide platforms for such a discourse.)[23] It may be important at this point to remind oneself that the history of artists in academic employment far precedes the debate on practice-as-research, and that many British universities have long had regulations for doctoral awards, for example, which allow for the submission of musical compositions or art exhibitions as PhD-level work. However, whilst traditionally such work would have been evaluated in terms of disciplinary mastery, within the framework of practice-as-research, performance practice is required to articulate itself according to scholarly parameters such as knowledge-making and knowledge-distribution, bringing it into closer contact with practice-based research procedures in fields outside the arts and humanities. Practice-as-research has become synonymous not with the introduction of creative methodologies into the researching of performance, which arguably have always been used in the field, but with the increasing recognition and institutional acceptance of such methods and the resulting need to establish protocols for evaluating them within existing academic structures.

The *PARIP–Practice as Research in Performance* project under the directorship of Baz Kershaw was established in 2001 (and ran until 2006) at the University of Bristol, with the aim 'to develop national frameworks for the encouragement of the highest standards in representing practical-creative research within academic contexts' (PARIP, 2001–06). Piccini and Kershaw (2003) have identified some of the issues that continue to animate the debate concerning this representation. These include:

- the attribution of 'professionalism' to certain artistic practices and their shifting relationship to definitions of 'academic' or academically produced practice;[24]
- a calling into question of traditional epistemologies based on a distinction between theory (as a critical and primarily cognitive activity) and practice (as a creative and embodied process), instead revealing the embodied, context-specific, and creative dimensions of theoretical research and the cognitive and critical aspects of practice;
- the question of whether a system of corroborating evidence is needed to differentiate the research dimension of practice from 'purely artistic' production;
- and what forms of dissemination (e.g., the document, the article, or the performance itself) may represent the particular knowledges produced by practice-as-research (connecting with familiar debates on documentation in performance studies).

In respect of the problem of dissemination, Piccini and Kershaw propose that 'an exclusive or over-reliance on outcomes other than the performance [...] crucially undermines some of the substantive philosophical underpinnings available to PAR [i.e., practice-as-research] [... as it] reproduces the systems of commodity exchange [...] implicitly critiqued by the rise of PAR itself' (2003: 122). At stake in this critique is, for the authors, nothing less than the scriptural economy that governs our academic system (at least in the arts and humanities) and thus the very identity of the university as we know it.

Piccini and Kershaw explicitly refer to performance studies as a discipline that is desiring and preparing for such a potentially fundamental epistemological change (Piccini and Kershaw, 2003: 113), citing Schechner's call for an integration of 'studying performance and doing performance' (Schechner, 2002: 1), Conquergood's attack on the academic 'apartheid of knowledges' (Conquergood, 1999, cited in Schechner, 2002: 18) that divides theory and practice, and Phelan's (1993) critique of knowledge reproduction as commodity exchange in the university. Phelan's own influential response to this critique has been the development of performative writing as a critical practice, which attends to the complex relationship between the occasion of performance and the writerly engagements which issue from it. Phelan has tied such writing closely to an ontology of performance as located in disappearance (Phelan, 1993: 146ff.) and to performance's affective force (ibid.: 11–12), both of which are re-enacted rather than represented in the performative event of writing.

In the United Kingdom, Adrian Heathfield in particular is continuing this project (Heathfield, 2006). For British scholar Susan Melrose, however, all writerly practices, whether more conventionally scholarly or performative, are reiterating the university's privileging of the scriptural economy. Melrose has for several years been developing a critical project which aims to rethink this economy within the context of the debate on practice-as-research. She argues that writing which engages with performance is usually limited to the particular 'perspective of expert spectating' (Melrose, 2006a: 120), and renames performance studies therefore provocatively as 'closet Spectator Studies' (Melrose, 2007). Performance's disappearance, which is central to Phelan's argument, is in Melrose's view specific to the experience of spectating 'and not at all appropriate to an understanding of performance practitioners' own "knowledge engagement" in performance production' (Melrose, 2006a: 121). The kinds of engagement that performance-makers are involved in during the process of creation are identified by her as frequently collaborative, mixed-mode, and 'looking forward with curiosity' (ibid., 126) to the appearance of something that the makers do not yet know, but know how to work towards. That Melrose expresses her critique of writing in a style which reproduces the retrospective, single-mode register of scholarly authorship does not devalue her point that performance practice from the perspective of the practitioners may be more appropriately theorized through forms of mixed-mode performance practices themselves.

Melrose also invites us to consider the particular knowledges that performance-makers draw on. She does so by attending to a notion that has always occupied an ambivalent place in the context of scholarly engagements with performance – that of 'disciplinary mastery,' or expertise. At the heart of the university's persistent anxiety over performance practice since the introduction of drama as an academic subject in the United Kingdom (see Shepherd and Wallis, 2004), the problematic of mastery, so Melrose proposes, has 'largely been erased from the discourses (and aspirations) of Performance Studies (while remaining central to ongoing and widely-celebrated activity in performance making)' (Melrose, 2006b: 132). She identifies a tendency in contemporary performance scholarship to privilege 'radical' artistic practice – which is often defined as a challenge to notions of mastery – and thereby to misrecognize the manner in which such radicality is in fact reliant upon a (masterful) execution of performance expertise. For Melrose, this misrecognition leads to a 'constitutive ambiguity' at the heart of Performance Studies, which produces 'exclusivist positioning [...] to such an extent that the study of performance (as distinct from Performance Studies) is split apart and fails to communicate internally' (ibid.: 134). There are doubtlessly certain critical fracture lines that run through the field and that create divisions between art and craft, innovation and mastery, scholarship and practice, 'text' and 'body,' writing and doing, the contemporary and the historical. Such fracture lines not only materialize as scholarly dissonances but also in the form of institutional structures (as in the division between universities and drama schools in the UK). But rather than bemoan it as a sign of disciplinary incoherence we may perhaps welcome it as an indication of the field's growth and diversification. Where it might become problematic in my opinion is in the field of performance pedagogy.

Practicing performance pedagogies

In a comparative analysis of a sample of performance studies programs at British and American universities,[25] Peter Harrop recently proposed that 'in the United Kingdom [...] the radical (for better or worse) may be lurking in the pedagogy' (Harrop, 2005: 196). Harrop here draws on already familiar models: he locates the 'radical' aspect of British performance pedagogy in its sustained engagement with and integration of creative practice, which has been central to the teaching of theatre and performance in Britain. For him, this centrality in turn 'decentralizes the role of language, particularly writing, as a means of control (or judgment) within the education system. For all its "trans," "inter," "anti" and "post"-disciplinary caterwaul the American model strikes me as a much more traditional and manageable academic discipline. It is people observing, talking and writing through a *theatrum mundi* panoptic in order to better understand the social construction of reality.' (ibid.: 199) There is, I would respond, nothing easily 'manageable' about the observing, talking, and writing of performance: to challenge an established bifurcation of teaching and research, to develop an interdisciplinary methodology, to recognize and harness the embodied nature of even the traditional classroom situation in order to foster experiential learning are central concerns of performance teachers in the United States (see, for example, the writings of Dolan, Stucky or Garoian) as well as in the United Kingdom (and elsewhere). Whilst not wanting to join Harrop's polemic, I do wish to take seriously his claim of the potential radicalism of a pedagogy that introduces students to different modes of knowledge-making through the engagement with and enacting of performance practice.

It may be useful at this point to take stock briefly of the current status quo of performance studies programs in the United Kingdom. Harrop observes that 'Performance Studies is not as omnipresent as I sometimes feel' (2005: 189), and indeed, there is actually little evidence in the United Kingdom for what Postlewait and Davis have identified as the 'growing popularity for jazzing up university theatre departments by renaming their components Performance Studies, whether or not they bear any relationship to the theoretical and methodological precepts of the foundational scholarship in performance studies' (Postlewait and Davies, 2004: 31). Although, as stated above, many drama and theatre departments in Britain 'do' performance studies in all but name,[26] it is interesting to note how few programs in this country in fact use the name. Of the six undergraduate degree programs currently offered at British universities that carry the phrase 'performance studies' in their title,[27] five use the phrase in conjunction with drama and/ or theatre studies, confirming the close association performance studies continues to have with its cognate disciplines. Only one program,[28] the BA degree at Aberystwyth University, on which I myself teach, carries the sole title 'Performance Studies.'[29] Among postgraduate programs the picture is similar. Harrop proposes that the naming of these programs is determined by two distinctive usages of 'performance studies' (ibid.: 190): one in which the term signals a reference to what Postlewait and Davis call the 'foundational' (i.e., US) scholarship in the field, in other words to a Schechnerian 'broad spectrum approach,' which seeks to analyze performance behavior across artistic practice, cultural performance, and everyday performativity; and a second one in which performance studies serves as a shorthand for a focus on contemporary performance practices, a privileging of the living avant-garde and innovative, interdisciplinary aesthetic practices (often leading inadvertently to reifying a canon of 'significant' artists and works). The latter use of the term, Harrop suggests, would therefore be better entitled 'Performance Art Studies' (ibid.: 193) (and indeed in a range of institutions is offered under titles such as 'new' or 'contemporary performance practices,' but also under the umbrella of 'theatre' or 'drama').

The two aspects Harrop has identified are not mutually exclusive – it is in fact their integration that probably best characterizes the pedagogy of performance (studies) in the United Kingdom, a pedagogy that has been largely developed in the context of undergraduate teaching (as opposed to the United States, where performance studies was first established and has primarily flourished as a graduate program). Whilst Schechner lays out performance studies' 'broad spectrum approach' as part of a call for a widening of the curricula in order to address what he perceives to be a conflation of 'the training of professionals and the education of majors' (Schechner, 1988: 4) in US performing arts departments, the educational context in which performance studies has taken hold in the United Kingdom is very different. The teaching of performance 'as a means of understanding historical, social and cultural processes' (Schechner, 1988: 6) that Schechner demands in place of skills training has arguably always been central to a predominantly research-led, non-vocational British university education in drama or theatre. The challenge that faces the teaching of performance studies here is an address to what Melrose has described as its 'constitutive ambiguity': to introduce students to and educate them in the conventions of their discipline whilst encouraging them to interrogate these conventions through a critical performative practice.

A closer look at our program at Aberystwyth University may make this clearer. The program attempts to take account of the particularity of its location in a small, predominantly Welsh-speaking town in rural West Wales by foregrounding the problematic of cultural context, identity, and representation. Whilst thus associating itself manifestly with the enquiries and approaches introduced by performance studies, the program focuses equally markedly on the teaching of contemporary, collaboratively conceived, innovative performance practices. Such practices have a strong tradition in Wales (and two of the teachers on the program, Mike Pearson and Jill Greenhalgh, have been instrumental in their development), but have increasingly come under threat through dwindling public arts funding (as a result of which Pearson and Greenhalgh have found themselves seeking academic employment). What the program aims to do thereby is to link certain compositional (and decompositional) procedures in performance practice to similar procedures in discursive-theoretical practice. It tries to encourage students to begin to reflect not just on knowledge itself (of whatever kind) but on the processes by which knowledge is made and expressed, and to engage with these processes creatively. To this end we have developed modes of assessment that include 'performed essays,' and forms of site writing that invite students to consider 'site as page' and 'page as site.'

Such educational aspirations, however, have to take on board the changing realities of the university sector in Britain, where on the one hand the Labour government is hoping to reach its target of accepting 50 percent of young people into Higher Education by 2010, which would make it the country's main provider of professional training, and on the other hand universities compete in a crowded marketplace over students who are beginning to assert themselves as fee-paying customers. As a result, the expectations of students of what a university education should deliver are often widely at odds with those of us teachers. The persistent – and in research terms largely productive – 'performance anxiety' which characterizes the practice of performance studies in the United Kingdom consequently often presents itself to the students as a problem which they must negotiate personally. Students are thereby not only required to reconcile their desire to acquire professional expertise (or, as Melrose would say, a certain mastery) in performance making with the challenge that the majority of theoretical discourses and creative practices they are being introduced to present to the very idea of mastery itself. They are increasingly also required to turn their practical and conceptual achievements in performance studies ultimately into a set of 'transferable skills' for a job in the new service- and experience-based economies.

The British way?

Baz Kershaw has recently proposed that the debate on practice-as-research presents 'a particularly lively UK contribution to the international growth of performance studies' (Kershaw, 2007: 51). Indeed, if there is currently such a thing as a recognizable 'British' approach to, or concern with, the study of performance we may with some justification locate it within this recognition of creative practice as a form of knowledge-making and knowledge-distribution,[30] even if its full implications are still being worked through. What may appear at first as a reversal of performance studies' focus from an extended consideration of cultural practices back to a narrow notion of performance as aesthetic production, in fact presents a profound re-evaluation of the nature of practice itself and our study of it. It would be naïve to suggest that this re-evaluation has been entirely driven by epistemological concerns. Whilst the rise of practice-as-research responds to the challenge to modernist traditions of knowledge-making that the techno-bureaucratic University of Excellence has ushered in, it also allows the university to draw capital from an expanded range of research activities and teaching opportunities. But whatever the motivation, if creative practice is to be fully integrated into the academy in the way it is being discussed and often already undertaken in the United Kingdom at present, it will have important consequences for our understanding of knowledge production and dissemination and thus for the practice of research and that of pedagogy in the field of performance studies.

Notes

1. Currently approximately £1.4 billion.
2. 1986, 1989, 1992, 1996, and 2001.
3. Other performance-oriented practices prevalent in UK Higher Education include annual staff appraisals, student evaluation of staff performance, regular quality assurance audits, quinquennial monitoring of departments, and the publication of league tables.
4. Up to four items of research output produced during the publication period (1 January 2001–31 December 2007) by individual researchers are ranked according to five quality levels: world-leading, internationally excellent, internationally recognized, nationally recognized, and unclassified (HEFCE et al., 2005: 31). Additional quantitative measurements are the institution's number of research students, research income, research environment, and indicators of esteem (HEFCE et al., 2005: 22–30).
5. The United Kingdom spends only around 1.8 percent of its GDP on research, which places it seventh among the *Group of Eight* leading industrialized nations, whilst it is second only to the US in the number of citations (Fazackerley, 2006).
6. Harley and Lowe are critical of the RAE's 'co-opting' of 'peer review for managerial ends,' as it 'offers individuals the possibility of securing material and symbolic rewards without ostensible violence to the traditional value systems which constitute academic identity' (Harley and Lowe, 1998: 24).
7. Documentation from earlier RAEs was not available.
8. In the context of what Schatzki, Knorr Cetina, and von Savigny have diagnosed as a 'practice turn in contemporary theory,' Knorr Cetina proposes that 'creative and constructive' research practice (as opposed to 'habitual and rule-governed,' which she identifies as 'performa*tive*' practice; my emphasis) is characterized by a continual reinvention of the way in which knowledge is acquired (Knorr Cetina, 2001: 175).
9. A number of competing terms are in use for this area of research. Kershaw offers the following distinction: 'practice-based' research refers to 'research through *live* performance practice, to determine how and what it may be contributing in the way of new knowledge or insights in fields other than performance,' whilst 'practice-as-research' is 'research into performance practice, to determine how that practice may be developing new insights into or knowledge about the forms, genres, uses, etc., of performance itself' (Kershaw, 2000: 138).
10. See Nelson and Andrews, 2003; and UK Council for Graduate Education, 1997.

11. My own location within this debate is as follows: educated in a German litera-ture department, I first encountered performance studies whilst undertaking my doctoral research in the UK. I was the first administrator of Performance Studies international and co-organizer of the fifth Performance Studies Con-ference in 1999, and now teach on the only dedicated performance studies undergraduate program in the UK at Aberystwyth University.

12. See the increasing number of publications that are devoted to such a 'return of theatre' 'from the heart of [...] the discourse of performance' (Ridout, 2006: 5) by UK-based scholars with an affinity to performance studies (Kelleher and Ridout, 2006; Kershaw, 2007; Ridout, 2006). Ridout locates the terms of this return in an 'identification of theatre with a certain kind of unease, and, in that unease, a possible "ontology" of theatre that might permit its reinstate-ment as a fruitful area of theoretical and political inquiry in spite of, if not because of, the case made against it or the alternatives to it offered by the discourses of performance' (2006: 7).

13. Reinelt is currently Professor at Warwick University.

14. Other UK-based publishers with a growing performance studies lists include Palgrave Macmillan and the university presses of Manchester, Cambridge, and Exeter.

15. Since 1995 the CPR is based at the University of Wales Aberystwyth (now Aberystwyth University).

16. Conferences considered performance's relationship with nature and culture (1989); politics and ideology (1990); ritual and shamanism (1993); process and documentation (1993); food and cookery (1994); tourism and identity (1996); places and pasts (1998) (Christie, Gough, and Watt, 2006).

17. PSi's presidency and administration has been hosted by UK-based universities for eight out of its 11-year existence to date (1998–2009): 1998–2001 Aberyst-wyth (President: Richard Gough); 2004–06 Nottingham Trent and 2007 Roehampton (President: Adrian Heathfield); 2007–cont. Warwick (President: Edward Scheer).

18. The PSi conference has since also been hosted by Queen Mary, University of London, in 2006.

19. *European Commission* website: http://ec.europa.eu/education/policies/educ/bologna/bologna_en.html (accessed 13 December 2008).

20. Other UK-based journals in the field include *Contemporary Theatre Review*, *New Theatre Quarterly*, and *Studies in Theatre and Performance*.

21. It cannot be reliably established how many departments currently teach drama, theatre, or performance programs in the UK. SCUDD, the Standing Conference of Drama Departments, which represents the interests of drama, theatre, and performing arts in the British Higher Education sector, currently has 75 member departments; see SCUDD website: http://www.scudd.org.uk/ (accessed 31 July 2008).

22. For a detailed history of practice-as-research, see Piccini (2003).

23. An exception here are the activities of the Live Art Development Agency (LADA), see: http://www.thisisliveart.co.uk/.

24. The attribute of 'professionalism' appears sometimes as an obstacle to and sometimes as the very foundation for research: The RAE 2001 report stated that 'the incorporation of practising professional artists into research cultures [...] sometimes [...] produces problems of integration with a consequent loss

of clarity in research aims and functions' (RAE, 2001). The 2008 RAE Guidelines, however, continue to define practice-based research as '[r]esearch in which knowledge is generated through *professional* practice' (HEFCE et al., 2004, inside back cover; my emphasis).

25. That is, the Master in Performance Studies at NYU (US); undergraduate provision in the Department of Theatre, Speech and Dance at Brown (US); the former BA Performance Studies (now BA Performance) at Northampton (UK); and the BA Performance Studies at Aberystwyth (UK).

26. In response to a call by Richard Schechner, the following universities in the UK identified themselves as working in the field of performance studies: Aberystwyth; Queen Mary London; de Montfort; Northampton; Bristol; and Roehampton (Schechner, 2006: 6–9).

27. See UCAS University and Colleges Admissions Service: http://www.ucas.ac.uk/.

28. Since Harrop undertook his survey, the University of Northampton, the first in the UK to introduce a BA in Performance Studies, has renamed its program 'Performance.'

29. The program is offered as a joint degree in combination with other subjects, including drama and theatre studies, international politics, English, or Welsh. Aberystwyth also offers a separate BA Astudiaethau Perfformio (Performance Studies) through the medium of Welsh.

30. See also parallel developments in Australia, Canada, Scandinavia, etc.

References

Allain, P. and J. Harvie. *The Routledge Companion to Theatre and Performance* (London and New York: Routledge, 2006).

Bacon, J. and F. Chamberlain. 'Editorial: The practice of Performance Studies in the United Kingdom,' *Studies in Theatre and Performance*, 25:3 (2005) 179–88.

Chamberlain, F. 'Interrogating boundaries/ respecting differences? The role of theatre within Performance Studies,' *Studies in Theatre and Performance*, 25:3 (2005) 263–70.

Christie, J., R. Gough, and D. Watt. (eds) *A Performance Cosmology: Testimony from the Future, Evidence of the Past* (London and New York: Routledge/Centre for Performance Research, 2006).

Fazackerley, A. 'UK Lags in Research Spending,' *Times Higher Education* (24 March 2006).

Harley, S. and P. Lowe. *Academics Divided: The Research Assessment Exercise and the Academic Labour Process* (Leicester: Leicester Business School Occasional Papers Series, 1998).

Harrop, P. 'What's in a name?,' *Studies in Theatre and Performance*, 25:3 (2005) 189–200.

Heathfield, A. 'Writing of the Event,' *A Performance Cosmology*, ed. J. Christie, R. Gough, and D. Watt (London and New York: Routledge/ CPR, 2006), pp. 179–82.

HEFCE. *Research Excellence Framework*, 1 July 2008: http://www.hefce.ac.uk/ Research/ref/ (accessed 10 July 2008).

HEFCE et al. *RAE 2001: A Guide to the 2001 Research Assessment Exercise* (Bristol: HEFCE, 2001a).

———. *RAE 2008: Initial Decisions by the UK Funding Bodies (Ref. RAE 01/2004)* (Bristol: HEFCE, 2004).

———. *RAE 2008: Guidance on Submissions (Ref. RAE 03/2005)* (Bristol: HEFCE, 2005).

Kelleher, J. and N. Ridout. (eds) *Contemporary Theatres in Europe: A Critical Companion* (London and New York: Routledge, 2006).

Kershaw, B. 'Performance, Memory, Heritage, History, Spectacle – *The Iron Ship*,' *Studies in Theatre and Performance*, 21:3 (2000) 132–49.

———. *Theatre Ecology: Environments and Performance Events* (Cambridge: Cambridge University Press, 2007).

Knorr Cetina, K. 'Objectual Practice,' *The Practice Turn in Contemporary Theory*, ed. T. R. Schatzki, K. Knorr Cetina, and E. von Savigny (London and New York: Routledge, 2000), pp. 175–88.

MacDonald, C. and R. Allsopp. 'The Temper of the Times,' *Performance Research*, 1:1 (1996) vi–viii.

Melrose, S. ' "Constitutive Ambiguities": Writing professional or expert performance practices, and the Théâtre du Soleil, Paris,' *Contemporary Theatres in Europe*, ed. J. Kelleher and N. Ridout (London and New York: Routledge, 2006a), pp. 120–35.

———. 'Who Knows – and who *cares* – about performance mastery (?),' *A Performance Cosmology*, ed. J. Christie, R. Gough, and D. Watt (London and New York: Routledge/CPR, 2006b), pp. 132–9.

———. 'Confessions of an Uneasy Expert Spectator,' *Professor S F Melrose* website: http://www.sfmelrose.u-net.com/ (July 2007) (accessed 1 August 2007).

Nelson, R. and S. Andrews. 'Regulations and protocols governing "Practice as Research" (PaR) in the performing arts in the UK leading to the award of PhD,' http://www.bris.ac.uk/parip/par_phd.htm (accessed 1 July 2007).

PARIP Practice as Research in Performance: 2001–2006 website: http://www.bristol.ac.uk/parip/ (accessed 1 July 2007).

Phelan, P. *Unmarked: The Politics of Performance* (London and New York: Routledge, 1993).

———. *Mourning Sex: Performing Public Memories* (London and New York: Routledge, 1997).

Piccini, A. 'An historiographic perspective on practice as research,' *Studies in Theatre and Performance*, 23:3 (2003) 191–207.

Piccini, A. and B. Kershaw. 'Practice as Research in Performance: From epistemology to evaluation,' *Journal of Media Practice*, 4:1 (2003) 113–23.

Postlewait, T. and T. C. Davis. 'Introduction,' *Theatricality*, ed. T. Postlewait and T. C. Davis (Cambridge: Cambridge University Press, 2004), pp. 1–39.

RAE 2001–UoA66 Drama, Dance and Performing Arts–Overview Report: http://195.194.167.103/overview/docs/UoA66.pdf (accessed 1 July 2007).

Readings, B. *The University in Ruins* (Cambridge, MA, and London: Harvard University Press, 1996).

Reinelt, J. 'Foreword from "across the pond",' *Contemporary Theatres in Europe*, ed. J. Kelleher and N. Ridout (London and New York: Routledge, 2006), pp. xiv–xviii.

Ridout, N. *Stage Fright, Animals, and Other Theatrical Problems* (Cambridge: Cambridge University Press, 2006).

Schechner, R. 'Performance Studies: The Broad Spectrum Approach' (TDR Comment), *The Drama Review (TDR)*, 32:3 (1988) 4–6.

———. 'A New Paradigm for Theatre in the Academy,' *The Drama Review (TDR)*, 36:4 (1992) 7–10.

———. *Performance Studies: An Introduction* (London and New York: Routledge, 2002).

———. *Performance Studies: An Introduction*, 2nd edn (London and New York: Routledge, 2006).

Shepherd, S. and M. Wallis. *Drama/ Theatre/ Performance* (London and New York: Routledge, 2004).

UFC, *Research Assessment Exercise 1992: The Outcome (Universities Funding Council Circular 26/29)* (Bristol: Universities Funding Council UFC, 1992).

UK Council for Graduate Education, *Practice-Based Doctorates in the Creative and Performance Arts and Design* (n.p.: UK Council for Graduate Education, 1997).

4

Rhetoric in Ruins: Performance Studies, Speech, and the 'Americanization' of the American University

Shannon Jackson

> The problem that students and teachers face is thus not so much the
> problem of what to believe as the problem of what kind of analysis of
> institutions will allow any belief to count for anything at all.
>
> (Readings, 1996: 192)

The posthumously published *The University in Ruins* was Bill Readings's attempt to reflect on the university's position as a social institution and occupant of an unstable zone in the increasingly globalized world of higher education. To reflect on the university in this way meant focusing one's attention less on the content and micro-moves within 'the culture wars' of the 1990s and more on analyzing the social, cultural, and institutional status of those debates themselves. Readings's book and other works have been helpful to me in my own attempts to come to terms with the institutional place of performance studies in higher education now.[1] My charge in this chapter is a little more precise for the purposes of this collection; it is to give both local and abstract accounts of some different kinds of performance studies pedagogy that developed in the United States throughout the twentieth century, practices signified by the gently mocked term 'oral interpretation,' whose history differs from the histories of PS that are most often told. In coming to terms with oral interpretation – what some call the 'Northwestern tradition' of American performance studies education, or what I and others have called the 'NCA tradition' – I will be tracking a rhetorical genealogy that simultaneously requires a rhetorical stance on itself. Indeed, it might well be this rhetorical stance – one that emphasizes not only *what* is valued in a field but *how* we do our valuing – that is the most vital element of the NCA tradition. Moreover, this emphasis on rhetorical performance might well be one

71

LIVERPOOL JOHN MOORES UNIVERSITY
LEARNING SERVICES

particularly vitalizing element offered by performance studies to what Readings called 'a post-historical university' more generally, vitalizing precisely because it feels itself to be dying.

Before recounting a history of oral interpretation as it changed and expanded in the twentieth century, a few notes about Readings's argument might be necessary in order to suggest how the notion of 'ruins' could signify the erosion of belief in performance studies even as it simultaneously renews a commitment to the field. Bill Readings's analysis offered its own take on an oft-recounted, but rarely internalized history of the academic profession – tracking its transforming processes of self-legitimation from a Kantian University of Reason to a Humboldtian University of Culture, and now to the techno-bureaucratic University of Excellence with which universities across the globe are becoming increasingly familiar. To academics used to thinking of themselves as perpetual 'resisters' of institutional structures, his arguments were soberingly inconvenient. Some of his most inconvenient claims were directed at a late-twentieth-century intellectual climate in the humanities that celebrated 'interdisciplinarity,' that fought the 'culture wars,' that propelled the 'rise of theory,' and that gathered around journals, centers, and occasionally societies that were affiliated, with varying degrees of commitment, with something called 'cultural studies.' While Readings's book was in sympathy with these and other movements that sought to diversify university curricula, he simultaneously noticed other unannounced premises and unintended consequences of their arguments. Most generally and reductively, Readings noticed the ways that such debates and transformations were not so much resistant to the techno-bureaucratic evolution of the overly managed university, but actually a symptom and, occasionally, a propeller of such administrative consolidation. As he and people like John Guillory noticed, the so-called 'culture wars' of the 1990s became as heated as they did precisely because a notion of culture had begun to matter less and less in United States society. Furthermore, if the Humboldtian notion of culture – in both its conservative and progressive orientations – was being eroded by techno-bureaucracy, then cultural studies' critique of culture only helped that process along. Indeed, Readings's argument was that a new University of Excellence was supported in large part by a process of 'de-referentialization,' one where there ceased to be any galvanizing principle or value beyond the leveling reductions of administrative accounting. While the exact implications of Readings's argument have been debated in several quarters, it has provided a set of questions that remain provocative and vexing for anyone trying to understand an interdisciplinary discipline. At the

very least, they ask us to think about how performance studies' own critiques of culture – however revolutionary or well-intended they might
have felt – might well be disconcertingly compatible with performance's
techno-bureaucratization as a discipline. Perhaps all of those claims
about performance studies as an 'anti-discipline' – claims that certainly
enticed my graduate student self in the 1990s – were actually symp-
toms of the de-referentializing impulses of a bureaucratic imaginary.
Indeed, the erosion of intellectual standards for assessing a discipline
unintentionally could allow accounting standards to enter in their place.

Readings's argument should obviously be important to a collection
such as *Contesting Performance* that seeks to combat the exclusively 'Americanized' story of the rise of performance studies. While, to my mind, the
term 'techno-bureaucracy' was under-theorized in Readings's book, not
always making distinctions between anti-governmental capitalist or pro-
government socialist versions of 'bureaucracy,' it does give us a way of
viewing some of the more insidious elements of academic employment
right now – whether the ideology of the 'bottom line' that animates more
universities in the United States or the 'Research Assessment Exercises'
endured by academics in the United Kingdom, Australia, New Zealand,
and Singapore. I would also submit that there is a particular generational consciousness exemplified in Bill Readings's book, one that is
symptomatic of many of us who were trained in the academy after its so-
called revolutions. There is something about being trained in the space
of erosion that seems to require an unusual amount of meta-disciplinary
reflection, something that academics used to do most often just before
retirement. The difference now is not only that meta-reflectors are getting younger (not just me, but also the editors of this collection), but that
a kind of critical suspicion overrides any articulation of such histories.
Ours is a generation who, in Marquard Smith's terms:

> is too young for punk and should have been too old for raves, their
> formative years lived with the threat of nuclear war at the forefronts of
> their nihilistic minds, with sexualities fashioned in a climate gripped
> by the fear of AIDS, and a political consciousness created wholly
> within and in opposition to the Thatcher-Reagan nexus, and thus
> attuned to both the consequences and pointlessness of organized
> politics.
>
> (Smith, 2005: 245)

The impulse to reflect about the profession does not come, then, from
the place it usually does, the place of nostalgia for what the university

was and the lament over what it has become. It is hard to be nostalgic if you cannot even pretend to claim access to a better Boomer past, whether 'better' is a university of the highest and noblest aspiration or 'better' is the activist university of a 1968 protest culture. Unable to narrate with the exquisite pain and exquisite pleasure of someone who was *there before now*, the task for many of us is to explain performance studies without ever having known anything else, to come to terms with what interdisciplinarity must be without really having known what a discipline was.

The narration of a rhetorical tradition of performance studies in the United States thus necessarily takes place within this awkward space of partial recall. However, there is also something uncanny about the fact that this tradition was so often 'unstoried,' to quote Paul Edwards (1999), even in the self-conscious debates about performance studies in the 1990s in the United States. It is thus strange to be writing this story of a US American genealogy in a book that simultaneously seeks to combat the Americanized story of the rise of performance studies. However, if we understand globalization to be wrapped up simultaneously with something that we loosely call Americanization, and if, furthermore, we understand the current state of universities across the globe to exist in a determining relationship with processes of Americanization, of which Readings's 'techno-bureaucratic University of Excellence' is an example, then I think that we need also to think about how US American disciplines and universities have also been Americanized. Indeed, paradoxical as it may seem, the internal regional and disciplinary debates about the 'two' strains of performance studies themselves exemplify the process of Americanization. The rhetorical genealogy of performance studies, including and especially its obfuscation, is itself a demonstration of Americanization in action in US America.

What is oral interpretation? A brief history

Answering the question 'What is oral interpretation?' might be just as hard as answering 'What is performance studies?' In this section, I want to offer a brief representation of the histories and experiences that collect around this rhetorical genealogy (Edwards, 1999). At the grandest, most Humboldtian level – one where the myopias of Western intellectual histories are perpetuated, even when they are never remembered very deeply – we can position rhetorical PS in the longest of classical genealogies, one that connects the act of orally performing a narrative to Western ancient traditions that are both rhetorical and poetic. This is to remember the Platonic and Aristotelian debates about the *rhetor*'s social function, techniques, and effects. This is to remember that the field of rhetoric has been historically linked to the goals of persuasion as well as to an analytic sensibility that understands how intimately the audience participates in the constitution of the *rhetor*'s expression. The more contemporary habit of understanding knowledge as socially constructed and as discursively produced is of course indebted to a longer strain of rhetorical thinking that has gone in and out of favor with changing intellectual trends. Meanwhile, another genealogy of rhetorical PS begins later but is still long, and it is the one that coincides with the concept of 'literature' as a category. As Paul Edwards rightly points out, the oral interpretation of literature as it became known in the nineteenth and twentieth centuries needed a concept of literature to become itself. As Margaret Robb has told the story, the oral interpretation of literature descended from a certain disciplinary moment at the turn of the twentieth century when the humanities fields that we now know as philosophy, literature, classics, or art history began to take distinct form (Robb, 1941, 1954). Coincident with what historians of the university call the transition from a classical to a vernacular curriculum, the study of literature became a distinct field that sought professional legitimation. Moving from the study of Shakespeare and classical texts to the inclusion of novels and then contemporary poetry, the field of literary studies formed. At the time of this syllabification, however, other elements of the nineteenth-century curriculum were gradually discarded or siphoned off to fields other than the literary. One of those elements was oral argumentation and performance, as both a form of 'public speaking' and as a hermeneutic form of oral poetics.

As I have argued elsewhere (Jackson, 2004), the performing professor was gradually defined as the opposite of what a growing literary professoriate felt itself to be, or felt that it needed to be in order to secure professional legitimacy for literary inquiry. As various groups formed and broke with each other in the history of the professionalization of literary studies – forming and re-forming organizations like the Modern Language Association, the National Council of the Teachers of English – teachers of oral performance and public speaking attached and detached themselves at various moments. As Paul Edwards describes it, one crucial moment occurred in 1914 when a small group of public speaking teachers, marginalized under 'Oral English' within the NCTE, 'decided to remain in a Chicago hotel until they could come up with a better idea. They emerged as charter members of the National Association of Academic Teachers of Public Speaking' (Edwards, 1999: 76).[2] This is the organization that relaunched the academic fields of rhetoric and speech communication as they are variously known in the United States today, eventually changing its name to the Speech Communication Association and most recently to the National Communication Association. And it is within this organization that the oral interpretation of literature would find its new home, casting off its delegitimized and feminized associations with elocution, and maintaining departmental wings and professional focus groups in the performance of literature, renamed the Interpretation Division in 1970 and renamed again as the Performance Studies Division in 1991.

Departments around the United States maintained this strain of the rhetoric and speech, hiring and reproducing curricula in the land-grant and state colleges that supported the field of communication studies most actively. Along the way, these professors of the oral interpretation of literature had to explain themselves continually to an increasingly 'scientizing' field of communication studies, arguing for the importance of literature to communication studies colleagues who were simultaneously developing the subfields in 'corporate communication' and 'organizational behavior.' When NCA changed the name of its Interpretation Division to Performance Studies, those communication studies departments – from North Carolina to Louisiana to Texas to Arizona to California to Minnesota to Illinois to Maine – followed suit. A Performance Studies curriculum existed and exists within schools of Speech or departments of Speech Communication in these and other states around the country. Northwestern University was one of the only places where an entire department – rather than a departmental wing – had been devoted to Interpretation, and hence its change to Performance Studies received the most attention, prompting some erroneously to assume that rhetorical performance studies was only a 'Northwestern' thing rather than a part of a national scholarly apparatus that had been in existence since 1914.

The rhetorical PS genealogy plots a slightly different story within the larger and uneven transitions from Humboldtian Universities of Culture to techno-bureaucratic Universities of Excellence. If the US American story of performance studies more often centers on the East coast, then theatre was the 'high cultural' site from which it claimed to be breaking. One way of telling this story institutionally is to see performance studies practitioners questioning the genre conventions and models of culture inherited from traditional disciplines of drama and theatre. To recount the genealogy of rhetorical performance studies, on the other hand, is to see similar and different connections to these intellectual trends and disciplinary dismantlings of the late-twentieth century. Oral interpretation had a delegitimated relation to literature departments, but not exactly the same delegitimated one that 'drama' professors had to the literary field. At the same time, this is also to tell a story about forms and practices that tried fervently to maintain a connection to a Humboldtian practice of culture, potentially doing so precisely because they had to survive within departments of communication studies whose increasingly social scientific fixations in 'organizational communication' already were allied with techno-bureaucratic definitions of knowledge. As oral interpretation professors tried to maintain their literary performance classrooms next to colleagues who were graphing the effectiveness of new corporate communication strategies, oral interpretation functioned as the last bastion of a University of Culture in departments that had already sold their souls to the University of Excellence.

What is oral interpretation?: a brief practicum

The practice of oral interpretation is based in the exploration and presentation of literature through techniques of performance. It is based in the belief that there is value in this process, a value that might be cast in critical, pedagogical, and artistic terms. In mid-century literary studies, it both resisted and reproduced New Critical values, positioning performance as a more public, potentially more unseemly form of literary exploration while simultaneously casting it as the ultimate form of textual close reading. Such a value translated to the pedagogical realm where professors of oral interpretation found their students more energized by, more engaged with, more discerning of the formal attributes of literary texts when such students took up the task of performing those texts in the classroom. Not only did the presumptuous act of performance force a minute attention on the innovations of an author's literary technique, the public presentation of those texts to a roomful of fellow students required a care and commitment that one could not fake. One might start with any piece of prose – say, Grace Paley's short story, 'Wants':

> I saw my ex-husband in the street. I was sitting on the steps of the new library.
> Hello, my life, I said. We had once been married for twenty-seven years, so I felt justified.
> He said, What? What life? No life of mine.
> I said, O.K. I don't argue when there's real disagreement. I got up and went into the library to see how much I owed them.

The librarian said $32 even and you've owed it for eighteen years. I didn't deny anything. Because I don't understand how time passes. I have had those books. I have often thought of them. The library is only two blocks away.

(Paley, 1994: 469)

This text goes on to exemplify much of what Grace Paley is known for – the incorporation of the mundanely domestic within the political, the simultaneous incarnation of a minute-to-minute temporality within larger chronological sweeps of time, the creation of a narrative voice that seems to be both inside and outside her life with an equal mix of awkwardness and erudition, a portrait of the poignant and painful operations of relationality that bring each of us, provisionally, into being. The premise of oral interpretation, its perpetually challenged hypothesis and hence animating principle, is that performing this text will make a reader more aware of these effects and the techniques that produce them. Such an act would have to consider acutely, for instance, when and where the speaker is at any moment, vacillating between what might be the scene of the story (at the library) and the scene of its telling (here in performance). Does the speaker say 'I saw my ex-husband in the street' to the audience? If so, at what point does she sit on the library steps to enter the past scene of the story? Before she tells us where she is? As she tells us? Or perhaps she enters the scene of the story only at the moment that she says 'Hello, my life.' A workshop of this text would probably try out all of these ideas and more. Along the way, participants might end up reckoning with the way that the text confounds neat temporal distinctions between such diegetic and extra-diegetic realms. Indeed, such a strategic conflation is arguably central to this text's effects, central to creating the sense of time as something to be endured and something to be recalled in strange and unanticipated shifts.

A slightly different, but related set of questions would come into play when we hear the response of the ex-husband: 'He said, What? What life? No life of mine.' The distinctiveness of this and so many other instances of reported speech in Paley's texts lies in the fact that she does not resort to quotation marks, or to what narratologists call 'direct discourse,' to represent it. The words of speakers and their interlocutors are given only modest separations of punctuation and indentation, making the decision about whether or not to turn this prose into a 'dialogue' an open one. A performer might change all registers of performance – focus, bodily comportment, and vocal tone – to create an entirely separate and continuous representation of a second character. 'What? What life? No life of mine,' this fully formed ex-husband might say back. But, taking a cue from Paley's own roundabout representation of speech, the performance of the line might try, not simply to represent a completely different character, but also to convey the effect of that character's speech on the ex-wife who narrates him: 'He said, What? What life?' she reports, performing the sounds of the exclaimed interrogative but also, possibly, the sting it attempts to instill. 'No life of mine,' his voice and hers continue together, one couched in the narrative of the other. The performance of these short phrases might thus attempt to use the simultaneity of their voices to perform the attachment of a detached relationality, conveying both the illocutionary intent of his dismissal and her attempt to refuse to be hurt by it. No life of mine, indeed. Are you following me? Probably not unless you try it yourself.

The passage continues, and the choices and experiments of oral interpretation would, too. I invite you to sound them out and try out different ones. The narratological experimentation (and the problem of re-presenting it) only become more complicated when oral interpretation moves into ensemble performance. Along the way, it is important to notice the narratological complexity of any choice. Whether playing with tense, pronoun, or physical and verbal discontinuity, the performer is often in several places at once. It is this kind of subtly unsettled position that Wallace Bacon (1979) called the 'tensiveness' of oral interpretation, a 'matching' between performance and text that never has the feel of perfectly sealed lamination. Not coincidentally, it was this kind of performance style that made Bertolt Brecht's concept of the 'not, but' (along with all of his other tense and pronoun exercises in 'A New Technique of Acting', 1992) much more comprehensible to me as a performer, in some ways restoring an actual narratological sense of what 'epic' acting should be. I am aware, however, that an illustration of oral interpretation might be challenging to read for a number of reasons. As a practice that exists in embodied, social, and temporally fleeting realm of a performance, it is a brand of performance studies research that appears awkwardly within the archival conventions of print representation. It shares what all field sites share and thus replicates what all fieldworkers experience when they become ethnographers – how exactly to represent the simultaneous experience of several registers adequately in the serial sentences of print? How to select words to represent the affect created or the gesture shared? The pedagogical and experimental aspect of the process, furthermore, is one whose significance lies as much in what is tried and tossed out as in what is tried and kept for a final performance. Indeed, a student performer's clarity about the manipulation of textual and performance techniques is something that emerges through repetitions, failures, alterations, and more repetitions. Thus, the elements that make a performance workshop seem so exciting are often exactly the elements that make a printed account of a performance workshop seem so boring. They also, not coincidentally, stubbornly refuse comprehensive 'accounting' in the spreadsheets and tallies that might legitimate the field in a University of Excellence. As a repertoire of vocal and gestural techniques that are perpetually revised and resituated, oral interpretation offers yet another illustration of the complex, asymmetrical relationship that Diana Taylor (2003) finds between the 'archives' and the 'repertoires' of performance research.

The other concern of course is that one minute example cannot give a sense of the vast diversity of performance texts encountered in rhetoric performance classes, especially as the late-twentieth-century scholars and activists expanded canons along gendered and cosmopolitan lines and as these and other scholars challenged the literariness of the text to include a whole variety of so-called 'non-fictional' texts. Before moving to analyses of one such expansion, it is worth noting at the same time how significant the oral interpretation brand of performance pedagogy and practice was for many artists in Chicago and beyond. Indeed, just as the 'East coast' genealogy has its artistic stars whose experiments are lauded and whose occasional brushes with mainstream success are discussed with mixtures of pride, suspicion, and jealousy, rhetoric PS has its stars, too. Mary Zimmerman or Frank Galati arguably function as its Spalding Gray or Liz LeCompte. David Schwimmer is its William Dafoe. If Lookingglass functions rhetorically as rhetorical PS's Wooster Group, then it is important simultaneously to notice that this line-up has its exclusions. Indeed, the range of influence extends to many artists and theatre groups, from directors such as Jessica Thebus, Eric Rosen, Martha Lavey, and Jim Lasko, to theatrical institutions such as About Face, Red Moon Puppet Theatre, and Steppenwolf. But the line-up also produces other kinds of exclusions, training attention on artists who have been influenced by rhetorical PS rather than other experimental companies; Goat Island paradoxically receives its most lavish attention from PS scholars outside of Chicago.

Nevertheless, a basic awareness of the art practices developed and revised in this milieu might help a reader and audience member rehear some of the Tony, Academy, Emmy, and MacArthur-'Genius' award-winning work that emerged from it. The adaptation-derived performance mode is behind Frank Galati's Broadway staging of *Grapes of Wrath* and the reason for his collaboration with Terrence McNally's adaptation of

Ragtime. It is behind his hiring and, we like to think, his merciful resignation from the *Seussical* musical. And it is the reason that Tony Kushner's *Homebody/Kabul* had its best staging at BAM under Galati's direction; it took a narratologist to figure out how to handle the abrupt change in tense, pronoun, and address between those two acts. Arguably, it also took a narratologist (and rhetorical PS artist), Mary Zimmerman, to figure out how to stage the stories of transformation, hubris, and loss in Ovid's *Metamorphoses* in New York in the year following 9/11. Rhetorical performance studies continues as one long experiment in the art and ethics of the addressive relation.

Expanding oral interpretation in performance studies

Not every example of every performance influenced by rhetorical PS suits everyone's taste or politics. And with that, it shares much with every other performance style one can possibly think of. Indeed, just as theatrical and other art movements altered contents and forms with changing political climates, so these techniques were revised and redeployed in a variety of academic and artistic contexts as the twentieth century wore on. If international PS scholars are familiar with any figure from the NCA tradition, that figure is most often Dwight Conquergood. Conquergood was an Assistant Professor of Interpretation at Northwestern University in the 1980s and helped to guide the name change to Performance Studies both in his department and in the subfield's division of the NCA. In 1995, he was a keynote speaker at the first annual meeting of what would become Performance Studies international. As the performance studies department chair in 1996, he hosted the second annual meeting of PSi at Northwestern and was the central master of ceremonies there. As I and others have argued, he is the person most consistently 'credited or blamed' (Jackson, 2006) with the move from oral interpretation to performance studies. Conquergood is known primarily for his ethnographic work in cross-cultural performance and particularly for the advocacy position he adopted on behalf of the people he studied and with whom he worked: Hmong refugees in Chicago and Laos, Latin King gang members in Chicago and elsewhere in the United States, death row inmates, and other disenfranchised groups in US society. To those more familiar with this work, it might be interesting to know that Conquergood's scholarly formation was quite solidly in oral interpretation. Indeed, just to reinforce a sense of alternate disciplinary histories, his dissertation used the figure of 'the Boast' in Anglo-Saxon England to investigate longer political histories around the division between literacy and orality.[3]

How exactly a scholar trained in an exceedingly 'Early' form of 'Oral English' became a performance studies scholar in contemporary ethnography is a question too large to receive adequate answer in this chapter.[4] But this kind of expansion certainly began to happen through a pedagogical route as much as any other when Assistant Professor Conquergood took over Northwestern University's course in 'Performance of Non-fiction.' There, propelled by a significant amount of post-doctoral re-skilling in courses with anthropologist Mary Douglas, Conquergood began to encourage students to conduct cross-cultural interviews as the basis for their performance work. The process of interviewing and reperforming an Other was thus positioned as a pedagogical means of confronting difference and defamiliarizing one's sense of self. Conquergood had himself participated in this brand of performance research in his own emergent work as an ethnographer of Hmong refugees in Chicago, re-presenting the voices of interview subjects who had fled invasion and encountered prejudice upon arrival in the United States. Most often, such performances were presented in community centers, courtrooms, or other civic sites where cross-cultural communication was necessary.

As graduate students at Northwestern, we all read an essay that was Conquergood's earliest attempt to come to terms in print with this change in his focus as a teacher and researcher, 'Performing as a Moral Act: Ethical Dimensions of the Ethnography of Performance' (Conquergood, 1985). Published in 1985 in *Literature in Performance*, the journal that would have changed its name to *Text and Performance Quarterly* by the time Conquergood published in it again, the essay was a self-conscious echo of Clifford Geertz's 'Thinking as a Moral Act: Ethical Dimensions of Anthropological Fieldwork in the New States' (Geertz, 2001). We taught it in every syllabus of our introductory courses in performance studies. Its appearance in syllabi in the 1990s did not go without a fight in the 1980s, however; we also all knew the story behind Northwestern's former Interpretation department chair, Lilla Heston. After reading the essay, she reportedly stomped down the hall to Conquergood's office and stood in his doorway as he watched her systematically rip it to shreds. Resistances notwithstanding, the essay was both symptom and propeller of the shift that brought the rhetorical practice of oral interpretation into the disciplinary formation of performance studies in the United States.

The touchstone for these courses and subfields in ethnographic performance derived in part from the fact that the social sciences were undergoing a period of self-reflection. In the 1970s and 1980s, there appeared a number of essays, collections, and books that investigated the status of the social science researcher, especially the ethnographic researcher in the midst of a field site where the imperative to 'write up' the site coexisted uneasily with the fact that that site was often opaque to the researcher thus charged. For Conquergood, and many other rhetorical PS scholars and students, the dilemmas of ethnography intimately paralleled the dilemmas of performance.

It is important to emphasize how much this expansion beyond the field of oral interpretation was simultaneously indebted to it. A match between ethnographic ethics and performance ethics could be imagined because 'performance' had been debated and practiced by these scholars as an act of translation and adaptation across worlds and across textual and embodied media for 75 years. While the world expanded and the politics of media were revealed in their complexity, Clifford Geertz's isolation of the ethnographic dilemma as a relation between 'being there' and 'being here' echoed the language of adaptation and its own preoccupations between the 'scene of the story' and the 'scene of the telling.' When James Clifford or Johannes Fabian critiqued the 'ethnographic present,' the mode of writing that sought to create cultural immediacy but that risked cultural typification, rhetorical PS practitioners who worried constantly about the effects of tense felt that they knew what they meant by 'chronopolitics' (Clifford, 1983; Fabian, 1983). And when Clifford (1988), and Marcus (Clifford and Marcus, 1986), and Geertz (1973 (2000)), and Rosaldo (1989) all experimented with the use of the first person, that is, their own first person, in an attempt to situate themselves as unstable researchers in a field site, rhetorical PS scholars were as attuned to the open vulnerability invited by a first-person narrator as they were to the fact that that same narratological choice could over-determine the ethnographic story.

The sense of performance not only as a narratological experiment but also as a cross-media experiment similarly supported the connection between rhetorical PS and ethnographic practice. The effort to represent the embodied, intimate, affective, ephemeral, perpetually revisable encounters of performance had been a source of institutional insecurity for performance, but it now had the possibility of generating an intellectually legitimizing opportunity for field practitioners. Indeed, a performer's mode of attention seemed exactly the mode of attention necessary to grapple with ethnographic dilemmas. When Edward Said critiqued the assumption that 'knowledge means rising above immediacy' (1979: 36), PS *rhetors* clapped. When anthropologist Michael Jackson said that 'textualism tends to ignore the flux of human interrelationships' (1989: 184), we cheered. And when anthropologist Talal Asad said that '[i]ndeed, it could be argued that translating an alien form of life, another culture, is not always done best through the representational discourse of ethnography, that under certain conditions a dramatic performance, the execution of a dance, or the playing of a piece of music might be more apt' (1986: 159), we thought that we had won the Lotto. The representational 'problems' of performance appeared for a while to be the representational 'solutions' to ethnography.

Conclusion

Whether all elements of the opportunity of performance studies have been actualized, this was a particular kind of conversation between ethnography and performance – one about narratological politics and about cross-media translation – that rhetorical PS offered and still offers to performance studies. It is important to notice that this particular conversation between performance and ethnography was as much or more about *how* to research than about *what* to research. While, for other types of PS scholars, the anthropological expansion of performance was about widening the objects of the canon to include performance forms from around the world, for many in the NCA tradition, performance ethnography was just as significantly an invitation to reflect about the politics and practices of exactly how to perform that expansion. It was and continues to be part of an ongoing investigation in the art and ethics of addressive relations.

I have come to realize, however, that the difficulty of writing this essay is partly about the difficulty of 'accounting' for an investigation and a genealogy that is so linked to pedagogy, where pedagogy's intimacies, its fleeting illuminations, its mindful interventions, its daily diligence, and moment-to-moment encounters are both intensely transformative and intensely undocumentable. Performance pedagogy is a *repertoire* if ever there was one, a space of human accountability that resists professionalized modes of accounting. In a move that might seem its own act of wishful thinking, however, I find myself wondering if it is exactly this obstacle to accounting in which we might find the potential for rhetorical performance in the ruined university. To think about rhetorical PS in ruins, then, is not only a naming of the fall of rhetoric, but actually a naming of the function of rhetoric in navigating higher education. My suggestion is made explicit in the figure to whom Readings turns at the end of his book. There, after 150 pages that read like conspiracy to many,

Readings turned to pedagogy amid the ruins, using language that managed to be committed and unbelieving at once. To value pedagogy was, for Readings, to value a site of social interdependence:

> In place of the lure of autonomy, of independence from all obligation, I want to insist that pedagogy is a relation, a network of obligation. In this sense, we might want to talk of the teacher as *rhetor* rather than as *magister*, one who speaks in a rhetorical context rather than one whose discourse is self-authorizing. The advantage here would be to recognize that the legitimation of the teacher's discourse is not immanent to that discourse but is always dependent, at least in part, on the rhetorical context of its reception. The *rhetor* is a speaker who take account of the audience, while the *magister* is indifferent to the specificity of her addressees.
>
> (Readings, 1996: 158)

What I find intriguing about Readings's elaborations of this rhetorical pedagogy is how much it mirrors the exchanges of the rhetorical performance classroom:

> If pedagogy is to pose a challenge to the ever-increasing bureaucratization of the University as a whole, it will need to de-center our vision of the education process, not merely adopt an oppositional stance in teaching. Only in this way can we hope to open up pedagogy, to lend it a temporality that resists commodification, by arguing that listening to Thought is not the spending of time in the production of an autonomous subject (even an oppositional one) or of an autonomous body of knowledge. Rather, to listen to Thought, to think beside each other and beside ourselves, is to explore an open network of obligations that keeps the question of meaning open as a locus of debate.
>
> (Readings, 1996: 164)

Whatever its limits, this is what the rhetorical performance classroom looks like. Performance practitioners of all varieties know the 'thinking together' that is rehearsal. Rhetorical PS adds to that interpersonal performance sphere a legacy of committed reflection on rhetorical contingency, about the politics of who is speaking, where and when, and about how those decisions betray subtle shifts of power that make different kinds of exchanges more or less possible. It is also medium-specific thinking together, one that does not fetishize the particularity of one real-time, co-present medium but that uses cross-medium acts of translation to foreground the dependence of thought on the medium of its enactment. The thing about performance pedagogy, too, is that it is also terribly inefficient, requiring enrollment limits that do not make financial sense, requiring extended hours that challenge the classroom schedulers. This brand of performance is based on a perpetually renewed space of obligation that simultaneously does not 'perform, or else' (McKenzie, 2001). Indeed, it is a brand of performance that refuses to be measured by the system of inputs and outputs that structure the 'performance evaluations' of academic departments with increasing frequency.

The trick for us now is to argue for the perpetuation of such a space while simultaneously knowing that it cannot be posited as a solution. 'Creating and addressing such an audience will not revitalize the University or solve all our problems,' says Bill Readings of a next generation of teachers; 'It will, however, allow the exploration of differences in ways that are liberating to the extent that they assume nothing in advance' (1996: 165). The generational task – and I do read the impulse to 'assume nothing in advance' as a symptom of a certain perspective learned and adopted by a generation of scholars raised in the context that I was – is to take our inheritances and figure out what to do with their oddities. It probably will not be to affirm the greatness of the literary writers as it once was or to perfect one's ability to 'know the Other' as might once have seemed possible. Oddly enough, it might be to notice the less legitimated aspects of our inheritance, since we come from senior scholars who never had the stature or institutional recognition that the 'fathers' of other disciplines once had. Our literature professors were too theatrical for literary studies; our communication professors were too cultural for communication science; our anthropology professors were accused of 'going native' in their political interventions and in their performing. The perpetuation of a relentlessly illegitimate, if undernoticed, discipline, for now over a century suggests that there are various ways of moving in and under the radar of whatever University (of Reason, Culture, or Excellence) one happens to be in. It can happen even in Americanized universities in US America. Working through this chapter – and, not coincidentally, cutting half of it in order to be 'accountable' to my word limit – has helped me to remember that largely unarchived network of practices and 'thinkings together' that remains central to my formation in the pedagogy of rhetorical performance studies, a network that has always been struggling to find itself and that, upon entering the performance classroom, provisionally always does.

Notes

1. For instance, Bourdieu, 1984; Graff, 1992; Guillory, 1993; Lyotard, 1994, to name a few.
2. While other scholars have told this story – Mary Margaret Robb, Mary Strine, Charlotte Lee, Robert Breen, Wallace Bacon, Beverly Whitaker Long, Mary Frances Hopkins, Linda Park-Fuller, Nathan Stuckey, Ron Pelias, Della Pollock, Sheron Dailey – Paul Edwards has done some of the most thorough recent remembering. Even more recently, Soyini Madison and Judith Hamera's edited collection (Madison and Hamera, 2006) represents a wider, longer, and more varied history of performance studies in the United States, including a section on 'Performance and Literature' and on 'Performance and Pedagogy,' umbrellas that allow the rhetorical PS genealogy a more secure footing in the histories of the field.
3. See a publication from that dissertation in Conquergood, 1983.
4. E. Patrick Johnson is working on a collection of Dwight Conquergood's many essay publications that will contribute significantly to the documentation of this important intellectual history.

References

Asad, T. 'The Concept of Cultural Translation in British Social Anthropology,' in *Writing Culture: The Poetics and Politics of Ethnography*, ed. J. Clifford and G. Marcus (Berkeley: University of California Press, 1986), pp. 141–64.

Bacon, W. A. *The Art of Interpretation*, 3rd edn (New York: Holt, Rinehart & Winston, 1979).

Bourdieu, P. *Homo Academicus* (Paris: Editions de Minuit, 1984).

Brecht, B. 'Short Description of a New Technique of Acting Which Produces an Alienation Effect,' in *Brecht on Theatre: The Development of an Aesthetic*, trans. J. Willett (New York: Hill & Wang, 1992), pp. 136–47.

Clifford, J. 'On Ethnographic Authority,' *Representations*, 1:2 (1983) 118–46.

———. *The Predicament of Culture: Twentieth-Century Ethnography, Literature, and Art* (Cambridge, MA: Harvard University Press, 1988).

Clifford, J. and G. Marcus. (eds) *Writing Culture: The Poetics and Politics of Ethnography* (Berkeley: University of California Press, 1986).

Conquergood, D. 'Literacy and Oral Performance in Anglo-Saxon England: Conflict and Confluence of Traditions,' in *Performance of Literature in Historical Perspective*, ed. D. W. Thompson (Lanham, MD: University Press of America, 1983), pp. 107–45.

———. 'Performing as a Moral Act: Ethical Dimensions of the Ethnography of Performance,' *Literature in Performance*, 5 (April 1985) 1–13.

Edwards, P. 'Unstoried: Teaching Literature in the Age of Performance Studies,' *Theatre Annual: A Journal of Performance Studies*, 52 (1999) 1–147.

Fabian, J. *Time and the Other: How Anthropology Makes its Object* (New York: Columbia University Press, 1983).

Geertz, C. 'Thick Description: Towards an Interpretive Theory of Culture,' in *Interpretation of Cultures: Selected Essays* (New York: Basic Books, 2000), pp. 3–32 (originally published 1973).

————. 'Thinking as a Moral Act: Ethical Dimensions of Anthropological Field-work in the New States,' in *Available Light: Anthropological Reflections on Philosophical Topics* (Princeton, NJ, and Oxford: Princeton University Press, 2001), pp. 21–41. (Originally published under the same title in *Antioch Review*, 28 (1968) 139–58).

Graff, G. *Beyond the Culture Wars: How Teaching the Conflicts Can Revitalize American Education* (New York: Norton, 1992).

Guillory, J. *Cultural Capital: The Problem of Literary Canon Formation* (Chicago: University of Chicago Press, 1993).

Jackson, M. *Paths Toward a Clearing: Radical Empiricism and Ethnographic Inquiry* (Bloomington: Indiana University Press, 1989).

Jackson, S. *Professing Performance: Theatre in the Academy from Philology to Performativity* (Cambridge and New York: Cambridge University Press, 2004).

————. 'Caravans Continued: In Memory of Dwight Conquergood,' *The Drama Review (TDR)*, 50: 1 (2006) 28–32.

Lyotard, J. F. *The Postmodern Condition: A Report on Knowledge*, trans. G. Bennington and B. Massumi (Minneapolis: University of Minneapolis Press, 1994).

Madison, S. and J. Hamera. *The SAGE Handbook of Performance Studies* (Thousand Oaks, CA: SAGE, 2006).

McKenzie, J. *Perform or Else: From Discipline to Performance* (London and New York: Routledge, 2001).

Paley, G. 'Wants,' in *You've Got to Read This: Contemporary American Writers Introduce Stories that Held Them in Awe*, ed. R. Hansen and J. Shepard (New York: HarperPerennial, 1994).

Readings, B. *The University in Ruins* (Cambridge, MA: Harvard University Press, 1996).

Robb, M. M. *Oral Interpretation of Literature in American Colleges and Universities: A Historical Study of Teaching Method* (New York: H. W. Wilson, 1941).

————. 'The Elocutionary Movement and Its Chief Figures,' in *The History of Speech Education in America*, ed. K. R. Wallace (New York: Appleton-Century-Crofts, 1954), pp. 178–201.

Rosaldo, R. *Culture and Truth: The Remaking of Social Analysis* (Boston, MA: Beacon Press, 1989).

Said, E. *Orientalism* (New York: Vintage Books, 1979).

Smith, M. 'Visual Studies, or the Ossification of Thought,' *The Journal of Visual Culture*, 4:2 (2005) 237–56.

Taylor, D. *The Archive and the Repertoire: Performing Cultural Memory in the Americas* (Durham, NC: Duke University Press, 2003).

5
Performance Studies in Japan

Uchino Tadashi and Takahashi Yuichiro

Japan is often cited as one of the places where performance art originated (Carlson, 2004; Goldberg, 2001). Despite the fact that such performance from the mid-1950s onwards – beginning with the Gutai group exhibitions and Hijikata Tatsumi's *Butoh* – has been documented by scholars and critics in Japan and abroad, there still has not been sufficient momentum in Japan itself for the study of performance to gain the status of an independent discipline. Within academia, performance has often been regarded with suspicion, and no performance studies departments have yet been established. Even the few theatre studies departments and programs in existence remain marginalized – and theatre scholarship itself is largely confined to dealing with literary texts.

As for performances studies scholarship proper, a handful of books on performance art, experimental theatre, and cultural performance have been published,[1] and oral interpretation has been advocated by a small number of US-educated scholars, often in conjunction with TESOL (Teaching English to Speakers of Other Languages) instruction. However, published translation of performance research in English is limited.[2] And while occasional courses on performance have been taught in departments of literature, culture, language, and fine arts, in comparison with the take-up of other recently developed fields such as cultural studies, postcolonial studies, or queer studies, academics in Japan have been slow to respond to performance studies.

In this chapter, Uchino Tadashi will examine the historical and cultural conditions that performance studies as a discipline has encountered in trying to enter the space of Japanese critical discourse, and Takahashi Yuichiro will discuss, in more concrete terms, what is visible as performance studies in Japan, and why this particular visibility has come about.

Why didn't performance studies become 'popular' in Japan?

Why didn't performance studies become 'popular' in Japan? I, Uchino Tadashi, pose this question not to unpack the question of agents and agencies – individuated or abstracted – involved in the historical process of appropriating performance studies in Japan, but instead to identify and outline the relationship between the pertinent sections of Japanese academia to the contemporary socio-cultural status of critical intellectual discourses that has resulted in the weak status of performance studies in Japan.

My hypothesis is that the development of performance studies in the United States and other places coincided with a moment in Japan when there was a rapid movement towards intellectual closure, and during which it was to be sociology that came to be considered the most important and relevant contemporary academic discipline. Increasingly in the 1990s, there appeared to be a felt need for a renewed and 'indigenous' critical but also 'national' language to speak about 'ourselves,' and sociology seemed to have become a *de facto* form of such an 'official' language and discourse both inside and outside academia.

This hypothesis is based on observation and personal interpretation, rather than on existing documentation or published research on the matter. In other words, nobody has made such a claim in an explicit manner. Some would disagree with this hypothesis, saying that the most important Euro-American critical writing in performance have been and still are being translated and published in Japan. But *how* the available material is used and circulated is also pertinent.

Journalism and academia at 'war'

I begin by making some observations about recent developments in literary criticism and theory in Japan, especially in the field of English literary studies, where the notion of performativity via the work of Judith Butler and Eve Kosofsky Sedgwick was welcomed, as least during a specific period. Who the readers are of such work and how the notion of 'performativity' has been deployed are the questions I am concerned with. Those observations that follow will lead to larger issues concerning the modernization of Japan since 1868, and the consequential project of dividing critical discourse between academia and journalism, especially as it came about in the critical discursive space of post-Second World War Japan.

There was an incident early in 2006 which may shed some light in an indicative manner on the matters addressed here. Judith Butler gave a lecture on 14 January 2006, at Ochanomizu Women's University in Tokyo. Some 850 academics and students are reported to have attended the lecture.[3] On the same day, there was a small public meeting and demonstration held at the Waseda University campus, not far away. The demonstrators opposed what they called the 'unjustified arrest' of a student who had distributed political leaflets to Waseda students in December 2005.[4] The student – who happened not to be from Waseda – was arrested by the police for tresspassing on private property. Only some 20 scholars, students, and sympathizers participated in that public meeting, versus the hundreds at Butler's lecture.

Suga Hidemi, one of the organizing participants of the Waseda meeting, commented bitingly on the situation, saying: 'We should question whether what guarantees the popularity of Butler's lecture is the fact that political issues are assimilated into research projects and academic curricula and thus have become depoliticized' (Suga, 2006: 21; translation Uchino). While there are other complications pertaining to this situation which cannot be dealt with here,[5] the central point I wish to make at this juncture is that I find Suga's anger understandable. Among the many 'imported' theoretical discourses in Japan, various forms of feminism have been exceptionally widely accepted, and have not only contributed to the reconfiguration of English literary studies in Japan, but have also been used to mobilize the formerly parochial sectors of activists, scholars, and policy-makers from various public spheres. However, at a critical moment – at least for Suga – when freedom of speech and expression were under threat, few feminist scholars were interested in supporting Suga and the demonstrators' cause at the nearby Waseda campus.

Suga is a literary critic, a member of the so-called '1968 generation,' and has been a major cultural activist for the last 35 years or so. Though he is currently a university professor, his main writing activity takes place in the realm of journalism. Suga's argument may remind us of important issues raised in a famous controversy between Slavoj Žižek and Judith Butler, published in *Contingency, Hegemony, Universality* (Butler, Laclau, and Žižek, 2000), regarding where the actual site of political struggle should be, and the ways in which theory can help in reconfiguring and redefining the political.

This brings us to the well-known and long-lasting disjuncture between academia and journalism in Japan's humanistic culture. The tradition of self-appointed, conscientious cultural critics who wish to participate in the political while staying outside of academia was firmly established, especially in the postwar years, when the roles academics and journalists are supposed to play in terms of intellectual commitments came to be relatively fixed.

To put it schematically, academia is the realm where people practice *gakumon*, and journalism is the realm where people practice criticism. *Gakumon*, according to *Kenkyusha's New Japanese-English Dictionary*, is: '(the pursuit of) learning; scholarship.' The term has other nuanced meanings, however, such as being alone in one's study, reading books and manuscripts. The term *gakumon* also implies scientific objectivity in pursuing 'the truth,' regardless of the use-value of the knowledge of that 'truth.' Especially in the humanities, where German philology has had a decisive influence on the institutionalization of scholarship in Japan, and in regard to the very understanding of what scholarship should 'do,' there developed deeply rooted assumptions about the object and the methodologies of scholarship – what academics study and how they are supposed to study their chosen objects.

I have been asked the same question again and again in various academic settings: how can you study theatre? After all, there is no 'fixed' text, so surely there is no fixed object of study? Even if one insists on something in one's paper or article, there is no way to objectively prove what you are saying is 'true.' The humanities in Japan, therefore, were – and basically still are – strictly empirical and historicist in orientation, especially in the study of things 'Japanese,' such as Japanese literature and history. And fortunately for such academic fields, there is more than 1000 years of scholarly history available to the present-day scholar, and also no lack of 'fixed' texts to study, as well.

In contrast to the development of academic criticism, literary criticism outside of academia in postwar Japan has witnessed the appearance of such influential figures as Kobayashi Hideo (1902–1983), Yoshimoto Takaaki (1924–), and, more recently, Karatani Kojin (1941–), and Suga Hidemi (1949–) himself, followed by a younger generation of critics who also remain outside of academia. As their general critical discourse was developed and defined against the empiricist-historicist traditions of academia, it was marked by the presence of personal sensibility and intuition, and sustained by abstract theory. Their work is targeted at the general reader, and is usually published by the commercial media, thereby making their critical interventions more visible and influential, suitable for the type and range of criticism being done. Thus arose the antagonistic relationship between academia and journalism, and the polarized objects of study split between them: academia is the realm for studying past cultural productions, while journalism is the realm for contemporary analysis and cultural praxis.

However, at least until the end of the 1980s, there was a productive and even healthy tension between journalism and academia in the humanities. The scholars exposed to Euro-American influences when they studied abroad could sustain a dialogical relationship with Japan's 'own' journalistic critical tradition, and this relationship was reciprocated. But with the burst of the so-called 'bubble economy' during the 1990s, which cultural critic Azuma Hideki calls the decade of the 'completion of [the] postmodernizing [of Japan]' (Azuma, 1999: 62–3; translation Uchino), a long-standing admiration for the West – so much a part of post-1868 Japanese history – seems to have vanished from the consciousness of the middle class. Modernization, taken as Westernization by then, seemed to be a concern of the past, as the West came to be so thoroughly internalized that the middle-class population became indifferent to the contemporary West. In place of the West as 'Other,' a renewed sense of 'Japanese' cultural-nationalist identity became called upon as a positive means to counter less-desired cultural identities; this has resulted, in one way or another, in a reaffirmation of a contemporary bourgeois identity mode that is indifferent to the West, given that bourgeois culture no longer seemed to suffer (post)colonial cultural cringe. This middle-class complacency – as it might be described – characterizes what I call a 'cultural closure,' which reinforced the aforementioned disjuncture between academia and journalism.

This 'great divide' between academism and journalism in the humanistic arena clearly militates against the possibility of intellectual cross-fertilization. The postwar period, and the last 20 years in particular, have seen the pronounced democratization of critical theory and discourse: who 'owns' theory and who speaks for whom became important questions for critics inside and outside academia. The 'great divide,' as a result, has been finalized. Whatever, say, Žižek writes is translated into Japanese almost immediately, while any book of importance in the field of performance studies is not translated so readily, as if to say Žižek or a writer in 'general' theory can offer more to contemporary Japanese cultural thinking than can any study and theory offered by performance studies. This is perhaps the result of the way that Žižek's work can enter critical journalism effectively, while scholars in performance studies still have to contend with the empiricist-humanistic (non-)orientation to performance.

Sociology as an integrating discourse and Japan's anti-theatrical tradition

It was sociology that came to bridge the gap between academia and journalism during the 1990s. That decade, in the middle of which Japan experienced such devastating events as the great Hanshin-Awaji earthquake, Aum Shinrikyo's sarin gas attack on the Tokyo subway, and the effects of the burst economic bubble, can be characterized as 'reflexive.' With the visible deterioration of older social relations, especially after 1995, people living in Japan seem to have felt the necessity to invent an indigenous critical 'national' language to speak about themselves. The necessity was manifested in one way as a popular front for a renewed sense of 'Japanese' national-culturalist identity. Sociology as a discourse seemed able to offer a critical language well able to engage with such identity concerns, both in journalistic and academic circles, and we witnessed a younger generation of sociologists such as Osawa Masachi (1958–), Miyadashi Shinji (1959–), and Kitada Akihiro (1971–) come onto the intellectual scene, publishing many books, both academic and journalistic in nature.

This, briefly, is the historical and cultural situation and context into which performance studies was introduced. The question then arises: why did sociology not show any interest in incorporating performance studies in inventing and developing their own language to speak about 'ourselves' in the 1990s? Here, we need to reflect upon the socio-cultural status of performance culture in Japan. It is not only because of the issue of 'the textual' that performance culture has not been the object of scholarship for a long time. We can sense, even now, deep-rooted anti-theatrical and therefore, by extension, anti-performance sentiments in Japan's cultural discourses. This is testified by the fact that in the process of modernization, theatre culture was never incorporated into the imperial and national project of institutionalizing various art forms. A national school of visual art, the Tokyo School of Fine Arts, was established in 1889, and a national school of music, the Tokyo School of Music, was established as early as 1879, at the dawn of Japan's modernization after the Meiji Restoration of 1868. Although there was an attempt, in the process of modernizing Japan's theatre culture, to establish a national theatre, theatre culture itself remained ensconced in the private and peripheral section of society. Even when the two national schools mentioned above were reorganized as the Tokyo University of Fine Arts in 1949, theatre was not added to the curriculum. To this date, no national university has a theatre studies department that is either academic or practical in its orientation. In the past, we did see a few theatre departments at private universities, though their number is increasing at the moment because of the changing climate surrounding university education in the last ten years.

Performance culture, for a long time, was regarded as a low, 'popular' art – for example, Kabuki was condemned as being pre-modern – while literature was accorded a 'national' status for its role in helping to conceive the new and 'modernizing' imagined community that emerged toward the end of the nineteenth century. Considering the speed at which Japan modernized itself, it is natural that literature, rather than theatre and its related genres of performing culture, became a privileged genre, as print culture can be more easily institutionalized and distributed in large quantities. Even in the postwar years, when Japan had to reconstruct itself as a renewed nation-state, literature remained a participant privileged genre in that process.

Performance culture in Japan also failed to acquire a viable relationship with other cultural genres, especially to literature. For someone to be an intellectual, whether in academia or journalism, came to mean – at least for our generation – that performance culture and its history need not be part of one's range of humanistic knowledge, although there was an exceptional decade that stands out – the 1960s. During that decade, Japan's performance culture became – perhaps for the first time in the history of modern Japan – an embodiment of the *zeitgeist*, when so-called *angura* or underground theatre practitioners, such as Kara Juro (1940–), Terayama Shuji (1935–1983), and Suzuki Tadashi (1939–) began to attract an audience influential in cultural circles. These practitioners' status, however, was ambiguous at best, as their work was discussed mostly in the journalistic arena, and theatre studies specialists had a hard time relating to *angura* theatre practitioners' experiments. To these specialists, *angura* praxis indicated a paradigm shift in thinking about theatre. This was not strictly accurate, as many *angura* practitioners claimed to either be revivifying or reinventing the tradition of Japan's physical theatre. However, the paradigm shift was considered to have occurred by those theatre scholars who concerned themselves with theatre taken only as textual – that is to say, with plays as written texts.

For journalism, however, there was no problem in incorporating the performative and physical aspects of theatre which *angura* theatre productions were problematizing. This did not lead to the academic theorizing and/or historicizing of *angura* theatre practices, but rather to the production of articles and books that stood side by side with the tradition of literary criticism in journalism that I have already discussed.

In short, theatre specialists were not able to 'reinvent' theatre studies to take into account performance and performativity. We did not have the equivalent of a Richard Schechner, who was both a scholar and practitioner, and who, in the 1980s, went on to experiment with and invent the notion of performance studies.

That performance studies, as a consequence of the above, did not have a chance to become widely recognized as an independent discipline is now a historico-cultural fact. This does not mean, however, that the issues performance studies raised never made it past what I take to be a form of 'cultural closure' that occurred. The notion of performativity, for instance, has been present, in varying degrees, among many different disciplinary discursive spaces for the last ten to 15 years, though many came to know the importance of the concept through – as I have mentioned earlier – Judith Butler's and Eve Kosofsky Sedgwick's work. This has to do with the fact that the 1990s was a decade when cultural studies as a tool of intellectual analysis was eagerly 'imported' through a variety of intellectual circuits, especially through sociology, which was gradually becoming dominant during that decade.

The traditional boundaries between sociology, literary studies, and other related disciplines, not unexpectedly, have been breached. Many of those who consider themselves to be 'cultural critics' take for granted the importance of 'performance' as a critical concept, as a genuinely political discourse, and as an object of study. Still, the pervasive anti-theatrical and anti-performative sentiments in Japanese academia remain, and work against a larger acceptance of performance studies and research – and I do not see these sentiments being easily overcome in the foreseeable future.

The great performance studies heist?

While I, Takahashi Yuichiro, concur with Uchino's analysis of historically adverse conditions that worked against the entry of performance studies into Japanese academia, I would like, in what follows, to illustrate that the study of, and research in, performance is thriving in Japan at a different locus – in the academy's periphery, at a galactic distance away from 'cultural criticism,' and in a manner sanctioned by the government that has promoted 'industry-academia cooperation' as a new model of researching knowledge and knowledge production aimed at invigorating the nation's economy as well as at the restructuring of (what some perceive as) ossified existing disciplinary formations. This policy was formulated in 1982 by the Science Council (*Gakujutsu-Shingikai*), an advisory body to the Minister of Education.[6] I am talking here about 'performance-*gaku*' (*gaku* in Japanese translates into both 'scholarship' and 'discipline'), a variant of performance studies developed and disseminated almost single-handedly by Sato Ayako (1947–). Sato is a professor in the Department of Drama in the College of Art at Nihon University, and she is also the Executive Board Chairperson of the International Performance Education Foundation. Her variety of performance studies, developed from its displaced US origins and transplanted to Japan, lacks what I take to be integral to performance studies: the counter-hegemonic critique of culture. Has she stolen performance studies from under our noses? Or could it be argued that the addition of this new dimension has enriched the field?

The fact that there has been a particular appropriation of performance studies, however, is not to say that the phrase 'performance studies' has no general currency. Translated as 'performance-*gaku*,'[7] it enjoys a fair amount of exposure to the Japanese public – thanks indeed, precisely, to the writing and innumerable lectures and television appearances of Sato. I would argue that her popularity reflects the current habits and various *habitus* of the Japanese in which the degree of (self-styled) cosmopolitan finesse has become a marker signifying class position. As her variety of performance studies can be viewed as an exemplary local (and indeed national) response to economic globalization, it merits close examination.[8]

Performance studies – Sato's variety

Sato Ayako claims that performance studies was launched in Japan in 1980, when she returned from the United States. Her point of departure was the concern voiced by Japan's business and political leaders that the Japanese, due to their characteristic shyness, had difficulties getting their message across in rapidly increasing cross-border and cross-cultural business and related negotiations. Globalization was beginning to be felt through mass air transportation, digital communication, and multinational enterprises, even though the term used in Japan was still the somewhat ethnocentric 'internationalization.' 'Internationalize, in order not to be left behind' had become a national rallying cry. Sato attributes the Japanese weakness to their cultural make-up. She explains that in a homogeneous, 'high-context' culture such as Japan's, communication often takes place without articulation. While Euro-Americans, brought up in a heterogeneous, 'low-context' culture, prefer explicit forms of expression, the Japanese, Sato maintains, rely on an implicit understanding mutually shared among insiders. Her answer to the national predicament was to develop a method to improve Japanese 'performance.' She defined performance as the 'conscious expression of self in everyday life,' slightly rewording Erving Goffman's formulation of the 'presentation of self' (see Goffman, 1959).

Although Sato was a student at New York University (NYU) when the Graduate Drama Department was transformed into the Department of Performance Studies, she does not subscribe to the liminal, and possibly subversive, model of performance that characterizes the work of Victor Turner and Richard Schechner. Her belief is that performance studies arose out of the demands that US society imposes upon individuals to effectively act out their social roles. She once remarked that in a society marked by individualism, meritocracy, litigation, and ethnic tensions, it is a necessity of life to be self-assertive: people need constantly to remind others who and what they are.[9] Thus, Sato argues, the shift from theatre to performance proved productive. She acknowledges Richard Schechner's contribution to performance studies as the introduction of theatrical frames to analyze performance in everyday life. In placing 'the expression of self' squarely at the center of everyday life, Sato relegates anthropology, theatre studies, and communication studies to the periphery of her research. Aesthetic and cultural performances, the staple objects of research in the United States, largely fall outside of her research agenda.

Sato's misreading of the NYU model of performance studies was, for her, a *felix culpa*. She was able to create her own variety, which she sought to make scientific and pragmatic – scientific in the sense of being objective and verifiable, and pragmatic in the sense of being pedagogically applicable. Sato, in coupling laboratory data with observation, has tried to measure and delineate characteristic Japanese behavior patterns. From there, she has gone on to develop curricula to teach the Japanese how to better express themselves. Performance is conceived as consciously organized behavior that is intended to make an impression on an audience through using skills that can be rehearsed and reproduced at will.

Sato bases her variety of performance studies on an emphasis of the body over text. According to her research, 70 percent of human performance is carried out non-verbally, while of the 30 percent that remains, 25 percent consists of para-linguistic use, and five percent of language use proper (Sato, 1995: 80). She therefore concludes that the most effective performance consists of a combination of non-verbal and para-linguistic skills. In her performance curricula, importance is attached to facial and bodily movements such as smiling, making eye-contact, and maintaining the proper distance between interlocutors. In the society of simulacra, it appears as if performance, or the expression that one simulates, dictates all. With her dictum, 'Ability not displayed is not ability at all,' Sato urges her fellow Japanese to acquire better performance skills.

Another important characteristic of Sato's variety of performance studies is her belief in the efficacy of performance as a tool for empowerment. She preaches that well-executed performance can guarantee better jobs, better friends, and, ultimately, personal happiness. Performative competence makes a learner motivated, confident, and likeable. The sense of achievement one attains is akin to Maslowian self-realization. At this stage, Sato's performance studies variant begins to approximate a form of self-enlightenment. Her theory of performance, which aspires toward science, simultaneously takes on a broadly 'spiritual' quality.[10]

To situate Sato's variety of performance studies in the context of globalization, it is necessary to note that it has appealed as strongly to businesses as it has to individuals. Empowerment experienced by individuals through performance is linked to enhancement of productivity and sales that can be achieved by businesses. For Sato, performance signifies both efficiency in the areas of work and socio-cultural behavior. Her research thus was embraced by Japan's service-oriented, late-capitalist market economy. She also established the International Performance Research Organization in Japan in 1992, the first academic society of its kind, in line with the industry-university cooperation model advocated by the government. In 1997, it became a foundation with a license granted by the Ministry of Education.[11] The foundation's webpage heading – 'why performative competence is effective'[12] – clearly calls out to business participation in its activities. The website enumerates the following benefits for companies that join the foundation:

1. [Performative competence] enables a swift response to globalization.
2. It makes communication easier within an organization.
3. It enhances the ability of a company to discharge information and boost sales.
4. It enables [employees] to scientifically assess client needs and respond in an appropriate manner.[13]

The website advertises that by acquiring performative competence, employees can communicate better and find meaning in their lives. Businesses, via Sato's prescriptions for competent performance, can improve their work environment, create a better public image, and make

operations more efficient. The benefits of performance extend in recip-
rocal fashion between life in general and business life. She reiterates that
the state, industry, and the individual must all become able performers
in the age of globalization.[14]

Performing Japanese ethnocentricism

Sato Ayako, while urging the Japanese to acquire performative competence on a par with Westerners, does not find fault or have any inherent
problem with modes of communication locally sedimented through centuries of practice in a high-context culture such as Japan's. Meaning,
she claims, must be explicit in the face of a global audience but can
be implicit in front of a local one. In a domestic situation, Sato's readers are advised to read the minds of their interlocutors, rather than
confront them verbally. 'How to express oneself' thus depends on a situation, on whether it is global or local. Sato's approach confirms that
the 'condition' of globality, while hegemonic, does not entail either
the renunciation of locality or necessarily subsume practices derived
from local knowledge. Sato recommends what she calls 'dual performance' – the ability to effectively code-switch between global standards
and Japanese particularity.

In one sense, Sato's performance pedagogy appears to open the way
for a multi-phrenic postmodern subject and subjectivity. Notions such
as 'plural identities' and 'flexible citizenship' can be associated with the
practice of dual performance. She seems to be sending out a message:
'Perform and play as many different roles as you can, and be effective.'
But when the issue of subjectivity is raised, we will see that Sato, working
within a state-industry-academy nexus, is establishmentarian and on the
side of the status quo. Her attitudes vis-à-vis nation, ethnicity, gender,
and class require critical scrutiny.

Dual performance posits a binary between globality (the West) and
locality (Japan). A facile stereotyping of oppositions conjectured between
them, such as general and particular, or low-context and high-context,
leads to the privileging of one term over the other. The assumption that
the Japanese share a high-context, homogeneous culture becomes unten-
able when the West is contrasted only with Japan. Sato comes precariously close to reductionism in presenting an essentialized Japanese-ness.
Placed in the context of the present revival of nationalism,[15] to speak
of the Japanese being uniformly shy, for example, is to totalize the concept of the nation. It embraces the discourse of a monolithic Japanese
identity which precludes the possibility of heterogeneous subjects and
subjectivity.

An example chosen by Sato to demonstrate para-linguistic differences by which speakers are distinguished is revealing. Citing an eyewitness account of a crime picked up by the local media, she writes: 'The muggers spoke in a fast, Southeast Asian[-type] English.' Sato explains that the way English was spoken could be identified by her immediately, given its regional peculiarities (1995: 43). Her remark, even if it is not intended, is insensitive. It reflects a type of sensibility that has been forged in Japan's modernizing process. What underlies Sato's remark is the 'othering' of the people of non-Japanese origin – here by associating Southeast Asians with crime; and by distinguishing them from the Japanese, it also makes a blanket statement about them.

The mindset revealed here is predicated upon two sets of binary oppositions, in which the former term in both cases is in hegemonic relationship to the latter term: the West and Japan, and Japan and (non-Japanese) Asia. Such a mindset also participates in a collective amnesia. It is forgetful of, first, the nineteenth-century socio-historical construction of Japan by the imperial government 'restored' in 1868 as a modern state, in which the Japanese came to be viewed as distinct from and superior to other Asian nations; second, Japan's colonization of Asia from the late-nineteenth century on; and third, the intra-regional cultural flows that have long existed in East and Southeast Asia.

Another example Sato raises to illustrate para-linguistic difference, which looks equally innocent at the surface level, illustrates her assumptions about gender and class. She relates an exchange between a company president (male, around 70 years' old) and a waiter (female, in her early twenties) that she has witnessed in a the lounge of an exclusive golf club (Sato, 1995: 117–18). The elderly male asks the younger female for a cigarette, which she brings to him. He then asks her for a cigarette lighter, which she also brings to him. Finally, he asks her for an ashtray. The president, when he made his initial request, expected the three items to be brought together. The waiter, on her part, was not aware of his implicit commands. The waiter is flatly described by Sato as a poor, inexperienced performer. Sato does not comment on the power matrix that has historically constituted Japan's social hierarchy. For Sato, a good performance is to behave according to one's station. The key to success is how well one can perform the role prescribed by society: a man, like a man; a woman, like a woman; a boss, like a boss; and a subordinate, like a subordinate. Sato's world is so clearly demarcated that it does not allow for an infiltration of heterogeneous space in-between existing formations.

Having examined Sato's adoption of the NYU model of performance studies, I have come to the conclusion that performance studies in the age of accelerating globalization can no longer be monopolized by those who profess anti-hegemonic positions. Sato's variety of performance studies illustrates that performance can be appropriated by the hegemonic, the majority, and the powers that be. Her use of performance as a pedagogic tool renders her approach normative and disciplinary. Students are empowered only when they subscribe to socially assigned roles and play them effectively. Beneath the liberal-sounding promise of self-enlightenment, performance as envisioned by Sato remains prescriptive and reinforces Japanese norms.

Sato's use of performance studies is also significant as a 'glocal' response to globalization. In advocating dual performance, she has indicated that globality and locality are not incompatible. She stresses that both global standards and local traditions must be performed with equal dexterity. I would like to point out, however, that globality and locality are both protean concepts. To postulate a binary relation between them, and to assume them to be stable and homogenous, is problematic.

Conclusion

The word 'performance' came into the Japanese vocabulary over the last 20 years or so. We hear the word every now and then on television or elsewhere in the popular media, where it almost always carries only negative and derogatory undertones. When people call some action a 'performance,' it is supposed to mean 'false,' 'deceptive,' and/or simply that the actor performing that action is shamelessly lying in some fashion. As Uchino has observed, this strangely limited usage of the notion of the transplanted word 'performance' is closely tied to the anti-theatrical tradition in Japan's intellectual and cultural history. Sato Ayako's culturalist intervention into this tradition and her naïve as well as reactionary theorization of the notion of 'performance,' as Takahashi sets it out, only perpetuates a negative image of performance studies within Japanese life.

The question, therefore, of whether performance studies is hegemonic, has not been and will not be an issue in the Japanese intellectual environment, because Sato's particular reactionary version, from its inception, is already hegemonic. Our roles as intellectuals with progressive commitments sharing a joint belief in the possibilities of performance studies, therefore, requires our participation in various forms of counter-discourses against Sato's popular version of performance studies. This, as is now patently clear, is what we have been trying to do in this chapter.

Notes

1. Representative works include: Ishi, 2003; Takahashi, 2005; Uchino, 2001.
2. Book-length translations are limited to: Benamou and Caramello, 1989 (1977); MacCaloon, 1988 (1984); and a selection of essays by Schechner, 1998.
3. 'Frontier of Gender Studies at Ochanomizu Women's University,' *Ochano-mizu University Institute for Gender Studies website*: http://www.igs.ocha.ac.jp/SITE1PUB/sun/9/news/report24.html?t= 1152879174080 (accessed 31 July 2006). The website is in Japanese only.
4. 'We Would Not Allow Unjust Arrest at the Department of Literature, Waseda University, on December 20, 2005': http://wasedadetaiho.web.fc2.com/ (accessed 30 July 2006). This website is only in Japanese.
5. Suga himself admits that 'I do not deny that an introduction of [various forms of] feminisms into Japan's academia has had an undeniable significance' (Suga, 2006: 21; translation Uchino); and that in terms of the politics surrounding 'gender troubles' issues in Japan, the question of who genuinely is politicized – or not – is a complex matter. As I note in the main text, a multifaceted feminism has helped to link together the formerly disparate groups of activists, scholars, and policy-makers. In short, 'style' in political resistance and intervention, along with the question of what constitutes the 'truly' political, would require a fully fledged discussion that is beyond the scope of this chapter.
6. The policy has been in force in various Liberal Democratic Party (LDP) administrations since Nakasone Yasuhiro became prime minister in 1983.
7. Because there is no Japanese equivalent word that can sufficiently convey the broad range of the English word 'performance,' the use of *katakana* transliteration is common. As for 'studies,' the term *kenkyu*, with its greater emphasis on 'research,' is used for fields such as cultural studies (*Bunka-Kenkyu*). The suffix *gaku* usually denotes more established disciplines such as philosophy (*Tetsu-Gaku*), linguistics (*Gengo-Gaku*), and biology (*Seibutsu-Gaku*). The distinction between the two terms, however, is not all that clear. Thus the use of *gaku* for performance studies cannot be ruled out as inappropriate.
8. Indeed, Sato's strategies make sense in what Baz Kershaw calls 'performative societies.' Kershaw points out that '[a]lthough the "performance" of companies, firms, shares, employees, institutions, etc., may be measured primarily in mundane material and/or statistical ways, the notion that they are "players" on an economic or industrial or civil "stage" is always implied by the usage'; he also observes that 'how individuals fare in [...] competition between life-styles or the struggle for survival depends increasingly on their [general] ability to "perform"' (Kershaw, 1999: 13).
9. Although Sato is a prolific writer, her ideas on performance studies are concisely summarized in *Jibun o Doh Hyogen Suruka: Performance-Gaku Nyumon* (*How to Express Oneself: Introduction to Performance Studies*; Sato, 1995) and in a reader she co-edited with Akiyama Hiroyuki, *Performance-Gaku* (Sato and Akiyama, 2001). This remark is taken from a roundtable discussion titled '*Performane ga naze gendai ni hitsuyo nanoka*' ('Why Performance is Needed Now'), that was attended by Sato Ayako, Akiyama Hirosuke, Konno Shozo, Ohshima Takeshi, and Takase Yoshimasa (ibid.: 10).

10. Sato is now no longer content with the definition of performance as 'the expression of self.' She thinks that performance must not merely be the sum of bodily techniques that can be rote learned. In the belief that performance must contain an essence that is inherently benign, Sato has redefined it as 'the expression of goodness of self' (Sato Ayako, 'Performance Studies: An Overview,' in Sato and Akiyama, 2001: 46).

11. The (International) Performance Education Foundation. I should like to keep the word 'international' in brackets as it appears only in the Foundation's English leaflet, and not in any Japanese texts approved by the government, without any intention to harp on Sato's propensity for slight exaggeration.

12. 'Kokusai Performance Kyoiku Kyokai' ('The International Performance Education Foundation'): http://www.spis.co.jp/ipef-entry.html (accessed 15 August 2003).

13. Translation Takahashi, as the website text is available only in Japanese.

14. Sato's vision has, in a way, already been realized by the state-directed economy of post-Second World War Japan, where the cooperation of workers through participation in quality control circles (QCC) was a significant support for management. In Sato's parlance, workers performed to improve corporate performance.

15. In the past two decades or so, despite protests made by neighboring countries, or perhaps because of them, nationalism in popular culture has been on the rise. Opinions on Japan's colonialism and past war crimes waver between self-indictment, willing amnesia, and self-justification. Those who propose that the 'natural' love of the country should be more freely expressed now seem to be ascendant.

References

Azuma, K. 'Tetteika-sareta Postmodan – Kyuju-nendi ni Tsuite' ('Completion of Postmodernity – About the 1990s'), *Musashino Bijutsu*, 111 (1999) 62–3.

Benamou, C. and C. Caramello (eds) *Postmodern no Performance* (*Performance in Postmodern Culture* [1977]), trans. T. Yamada and Y. Nagata (Tokyo: Kokubunsha, 1989).

Butler, J., E., E. Laclau and S. Žižek. *Contingency, Hegemony, Universality: Contemporary Dialogues on the Left* (London: Verso, 2000).

Carlson, M. *Performance: A Critical Introduction*, 2nd edn (London and New York: Routledge, 2004).

Goffman, I. *The Presentation of the Self in Everyday Life* (Garden City, NY: Doubleday, 1959).

Goldberg, R. L. *Performance Art: From Futurism to the Present*, rev. and expanded edn (London: Thames & Hudson, 2001).

Ishi, T. *Iso-no-Sexuality* (*Transvestite Sexuality*), 2nd edn (Tokyo: Shinjuku Shobo, 2003).

Kershaw, B. *The Radical in Performance: Between Brecht and Baudrillard* (London and New York: Routledge, 1999).

MacCaloon, J. J. *Sekai o Utusu Kagami* (*Rite, Drama, Festival: Rehearsals Toward a Theory of Culture* [1984]), trans. H. Takayama et al. (Tokyo: Heibonsha, 1988).

Sato, A. *Jibun o Doh Hyogen Suruka: Performance-Gaku Nyumon* (*How to Express Oneself: Introduction to Performance Studies*) (Tokyo: Kodansha, 1995).

Sato, A. and H. Akiyama (eds) *Performance-Gaku* (Tokyo: Shibun-doh, 2001).

Schechner, R. *Performance-Kenkyu* (*Performance Studies*), complied and trans. Y. Takahashi (Kyoto: Jinbunshoin, 1998).

Suga, H. 'Post-Jichi Kukan – 2005-nen 12-gatsu 20-nichi Waseda Daigaku ni okeru Billa-maki Taiho wo Megutte' ('Post-Governing Space – On the Arrest of Handing Out of Leaflets at Waseda University Campus on December 20, 2005'), in *Neolibeka-suru Kokyo-ken* (*Neoliberalizing the Public Domain*) (Tokyo: Akashi Shoten, 2006) 7–24.

Takahashi, Y. *Shintaika Sareru Chi* (*The Embodiment of Knowledge*) (Tokyo: Serika Shobo, 2005).

Uchino, T. *Melodrama kara Performance* (*From Melodrama to Performance*) (Tokyo: Tokyo University Press, 2001).

Part II
Contesting the Academic Discipline through Performance

6
Between Antipodality and Relational Performance: Performance Studies in Australia

Edward Scheer and Peter Eckersall

> Let's make a theory of performance collapse!
> John Forbes, *'Satori in Viterbo'* (1998: 21)

I like America/America likes me?

On Friday 30 August, 2002, a Performance Studies (PS) Symposium held at the University of New South Wales (UNSW) in Sydney addressed a variety of questions about the 'current disciplinary status of and the scope of research in Performance Studies in Australia' and 'whether there is such a thing as "Australian Performance Studies" and, if so, how it differs from Performance Studies in other geographical locations, and especially from the United States.' These questions was framed by the principal organizer of the event, Dr Moe Meyer, and the discussion chaired by Dr Sharon Mazer, both Americans working in Australia and New Zealand respectively. The responses were as varied as one might expect from scholars whose disciplinary backgrounds and institutional affiliations differed considerably and whose research took them across cultural studies, popular culture studies, theatre studies, and the visual arts, just to list a few of the key areas which were identified.[1]

One fairly prominent strand of response resisted the attempt to define an Australian version of PS, arguing for an open-ended sense of the field as practiced in Australia. This response seemed at least in part to be inspired by a form of *ressentiment*, as if the question were an attempt by American colleagues to perform a 'quick and dirty' variety of ethnography on the culture of Australian PS scholars. One of the delegates, Ian Maxwell (of the 'quick and dirty' reference), described the discussion at the symposium in terms of 'a struggle involving a rather uncomfortable transposing of the habitus of one local academic field to another relatively autonomous local academic field. The struggle to legitimize and institutionalize a particular disciplinary center in this instance was enacted precisely as disciplinary action, complete with threatened sanction: you will not be/are not recognized and are at best marginal' (Maxwell, 2006: 36). The question itself then became readable as a pseudo-colonizing gesture and not an act of critical inquiry.

This was not an isolated instance of this kind of response. In 2000, *Australasian Drama Studies* (*ADS*), one of the journals of the Australasian Association for Theatre, Drama, and Performance Studies (ADSA), the peak body representing performance studies and theatre studies scholarship in Australia, produced a special issue on PS in Australia.[2] This issue is replete with critical commentary on the American hegemony of PS in a number of articles by scholars such as Gay McCauley, Glen d'Cruz, and Rachel Fensham among others. Indeed, for anyone with a passing knowledge of the history of discussions about the place of PS in Australia, this corresponds to a pattern of response which appears at first glance to be inspired by a discourse of anti-Americanism.

Accounting for this in the specific context of PS entails dealing with three related problems: first, the discourses and practices of PS have been developed and promulgated largely by North American scholars in places like New York University and Northwestern University; second, their application in Australia is problematic precisely because of their usefulness and potency as discourses complicated by contemporary American cultural and political hegemony; and third, due to the peculiar sensitivities in Australia to colonialisms of any kind. This last point will be the focus of the discussion which follows.

In Australia, PS scholars tend to wear our postcolonial hearts on our sleeves. The oppositionality this produces is not therefore reducible to knee-jerk anti-Americanism, but is due to a broader suspicion of colonizing gestures and a heightened sensitivity to forms of discourse which might in subtle ways replicate these gestures.

Ian Maxwell ends his description of the symposium by adding a telling addendum: '(Of course, we are, as antipodeans, always already marginalized, so this comes as no surprise.)' (2006: 36). This is meant as a throwaway line, but it encodes a basic aspect of the discussion which no-one in this field has yet taken up, as it raises some key questions not only about PS in Australia and who is doing it, but about the nature of the very marginalized 'antipodeans' of Maxwell's phrase. His assertion is that Australia's marginalization is an effect of its antipodality, of its geographical oppositeness, and therefore it should come as no surprise, since we expect it and perhaps occasionally strategically reproduce it. This kind of oppositional ressentiment is not incidental to the local habitus in PS and elsewhere. In fact, one could argue that it is a key performative element of the Australian cultural context. It is performative in Judith Butler's sense of that term, as a constitutive but not essentially natural gesture at the level of the signifier of Australian-ness. So before any analysis can take place about the specific place of PS in Australia it is important to step back and provide an account for this aspect of the discursive construction of Australia, beginning with the very sensitivity provoked by the demands from American colleagues at the symposium.

This particular sensitivity requires analysis informed by, but not limited to, the now-familiar critical frameworks of postcolonial studies. Most PS scholars in Australia are white and European. There is a reluctance to address the positioning of white and European perspectives within Australian society which may disguise other forms of hegemony, the way that whiteness, as scholars such as Richard Dyer (1997) and Ghassan Hage (1998) have argued, colonizes representations of ordinary otherness. So non-whites have to perform a representative function that most whites are literally oblivious to even as we require it of non-whites. The fact is that PS in Australia is almost an exclusively white PS. It is the study of European forms of culture and behavior with a largely tokenistic account of non-white or non-European cultures. This is of course not a problem that is exclusive to PS – but PS is precisely the field in which such questions should be raised and not dismissed as a colonizing gesture, no matter how aggressively framed.

In a way, postcolonial studies provides a ready-made answer to these questions in terms of the ineluctable impact of the historic patterns of imperialism on host cultures and societies, which then transact their own cultural business in similar ways. But there is a need to move beyond the terms of postcolonialisms and into conversations that, however inflected with the voices of empire, can attempt to speak in other ways to other cultures. To attempt an alternative account of Australian oppositionality is an important task which may have implications for a broader analysis of the complexities of cross-cultural research in which the assumption of Australia as a benign and curious fellow traveller with the cultures of the region might need to be revisited, in terms of how we are seen by them as well as how we see ourselves. This is especially important in the recent political climate in which Australia has actively promoted itself as America's deputy in the Pacific.

The *ADS* issue which surveyed the field in Australia and provoked these questions as to the Australianness or otherwise of PS in Australia did not attempt to answer them. The editors took an institutional perspective, beginning with Sydney University's center for PS with its background in the ethnography of theatre practices and theatre semiotics, and its newly developed breadth of approach encompassing a broad range of meaningful cultural acts. This is in some ways emblematic of the journey of PS in Australia, which continues to move away from an anthropology of theatre towards a theatrical anthropology of cultural and art performances as varied as music, drama, ritual studies, and sport. The *ADS* issue was especially useful in charting different methods in the pedagogy of PS in Australia, but it did not attempt to theorize how the local performance of culture, and the research it generates, can be connected with larger questions about the production of Australian identity. Nor did it address the construction and performance of Australianness and otherness, in the intercultural dynamics of Australian experience.

These are enormous topics, beyond the scope of this chapter, but it is important that we begin to address them. In the present argument, we propose a sketch of such an approach in two stages: first, to attempt more clearly to frame the topic of Australian culture in its historicity and its regional contemporaneity; and second, to then facilitate the analysis of the performance culture embedded within that larger entity with the objects and tools of international/American performance studies in mind, and perhaps then to address the questions raised in the 2002 symposium. To deal with the first part, we return to Maxwell's remark about 'we antipodeans' and tease this term out a bit, and for the second, we will present a brief case study of a project which enables some key emergent themes of PS in Australia to be identified.

Antipodality

In the 1980s, Australian cultural studies and visual arts scholars, especially those associated with the art journal *Art and Text*, began to address these very issues of Australian identity, and its relations with Europe and America as opposed to Asia, that by 2002 were surfacing in PS. One of these discussions concerned the notion of antipodality, an idea which has an intriguing and important history in Australian studies as a discourse predicated on the notion of the antipodes, the sense from both North and South of a distant opposite earth where things really should work very differently. It is a useful discourse for the kind of oppositional response associated with the practice of PS in Australia because it allows for a discussion of the differences internal to Australian society, while also accounting for the reasons why Australia refuses to live up to the expectations of its own 'difference' – its national mythology and a singularity imposed from both within and without.

The term 'antipodes' – literally 'having the feet opposite,' indicating those who dwell directly opposite each other on the globe – has always provoked a certain amount of ambiguity in the discourses of the human sciences. Plato's *Timaeus* affirms the sphericity of the earth but stresses that cosmographical descriptions implying opposition – that a point is 'above' or 'below' another – inaccurately reflect this sphericity (1970: 268). In Plato's day the expression 'down under' would have been anathema, much as it is now to many of us.

Cosmographic inconsistencies aside, the concept of the antipodes has evolved in the discourses of Australian culture and history to designate an Australia as seen from the parent cultures and economies of Europe. At the other end of the world, Australia was constituted as a nation under the aegis of this term and the perspective it both signifies and engenders. Now over 100 years after federation, the post-Second World War policy of 'multiculturalism' has produced an Australia primarily constituted by clusters of migrant peoples of wildly divergent beliefs and practices, united only by what the European émigré writer Emile Cioran described as 'a nostalgia for space, a horror of home, a vagabond dream and a need to die far away' (1987: 61). In this perverse context 'multiculturalism' describes a cohering policy agenda which also serves to reinforce those anterior antipodal claims. Ironically, it is a term which has recently come under attack from neo-conservatives in the government as not producing the right image of Australia as an Anglotopia.

The markedly European flavor of much performance culture in Australia, disavowing both regional geography and indigenous ontology, is sometimes surprising for non-Australians anticipating a more functional antipodality, a more pronounced sense of difference. Australian culture thereby becomes cast as 'the flak of an explosion not of our detonation' (Paul Taylor, cited in Morris, 1983: 5). But the 'our' in Paul Taylor's apt phrasing is also somewhat problematic, for the concept of the 'Antipodes' has not served to homogenize an Australian cultural identity. This alone might account for some of the difficulty the questions of the 2002 symposium posed to Australian scholars.

In this context, Australian culture becomes at worst a kind of dumping ground for jettisoned European or American product, at best an anamorphosis of the 'parent' cultures. The antipodes, as Morris has suggested (1983), serves to both guarantee the authenticity of England and America as points of origin for Australian cultural practices, while yet acting to contest it.

So 'antipodes' signifies, in the realm of culture, the locus of responses to the normative predicates of the 'parent' cultures: alternatives as well as replications, both of which bind the production of Australianness to these origins. Traces of the former are everywhere in the affirmations of identity usually associated with sporting moments (for example, the 2000 Sydney Olympics) and patriotic support of national teams (Australia's powerful cricket team and less dominant rugby side, the Wallabies). Replication cultures subsist in the antipodal soil in the form of nostalgia, such as the pro-monarchy movement in Australia which defeated the referendum on an Australian republic in 1998. Both these options tie Australian to English origins. The Olympics had to be opened by the head of state, the Queen of England, or her representative, the governor general, while rugby and cricket are both creations of English cultural formations.

At the same time, the term also articulates an internal sense of otherness. With the bulk of the population inhabiting the coastal fringes, the outback had become mythologized and indeed antipodalized as the red center, the dead heart, the amorphous 'interior,' an uninhabitable nowhere, where 'birds fly backwards, rivers run against nature, the sand spawns fish' (Morris, 1982: 70) and which might as well be put to good use as a uranium quarry, nuclear test site, or as a strategic information gathering site for the American military. For cultural critic Meaghan Morris, antipodality explains the phenomenon of cultural cringe, that 'if much Australian cultural activity today is engaged in a radical forgetting (rather than violation or contestation) of original codes, then it is by virtue of the recognition that a compilation culture of borrowed fragments, stray reproductions, and alien(ated) memories is what we already have to begin with' (1983: 7).

Yet the undermining of Australia as alternative – which for Morris, writing a generation ago, was the cohering force in Australian cultural analysis – is still a largely negative movement of disavowal (of being 'down under,' and so forth), a repudiation of Australia as a site of cultural difference to European and American cultural centers. In this sense the durability of this discourse is unremarkable in that even the baldest affirmation of Australian uniqueness (as opposed to other models) must always refer to that condition which it rejects or seeks to destabilize. 'Antipodality' collects those acts of signifying otherness which have become internal to the structuring of an Australian culture and divides European from indigenous perspectives to this day.

Relational performance: PS and the 'region'

If there are Australias beyond the antipodes, so that antipodality would no longer be the key motif for an Australian cultural analysis, it is perhaps to be found not in cultural origins, but in present-day practices of regional interactions and engagement, in modes of production which neither seek to affirm nor negate a particular characterization of Australianness. This kind of relational approach, rather than an approach seeking to reify the location of Australian culture and identity, can be found in the recent history of experimental performance practice in Australia. Such an approach, in which an intentional suspension of national and cultural identity is undertaken as part of the project, is rare in the history of more mainstream cultural practices in Australia for obvious reasons to do with the essentially antipodal forms of, for instance, drama or opera inherited by the colony before it formed into the Federated States of Australia. There is less room in more traditional forms for suspension of identity, as a reified identity is encoded within the form itself.

Some of the forerunners to PS adherents are an interesting study in this regard. Artists such as Lindzee Smith at the Australian Performing Group (that alongside La Mama Theatre was home to nascent performance art and new-wave theatre movements) sought out the underground Japanese theatre of Terayama Shûji and Kara Jûrô. Interest in Eastern performance cultures was a counter-cultural action, likely coupled with opposition to the Vietnam War, Maoism, and the New Left. The late-1970s saw interest in the work of Javanese playwright-director Rendra, whose *Struggle of the Naga Tribe*, combining *wayang kulit* and anti-capitalist political commentary, was translated by Max Lane in 1980. There are numerous examples confirming the general point that for many Australian artists, regional engagements with colleagues in the near neighborhood revolved around notions of international solidarity and resistance to dominant cultural systems, the problematic US-Australia nexus included. Even so, the extent to which this interest arose from the extended coverage of Asian performance practices in journals such as *The Drama Review* (*TDR*) cannot be measured and complicates our historical understanding of PS, especially as an emergent field in the 1970s. The distancing of Australian performance from Euro-American trends also involved the interpretation of non-Western practice through the avant-garde 'intercultural' practices of Jerzy Grotowski and Peter Brook, but remains largely inchoate.

The emergence of 'regional,' mostly corporeal vocabularies of exchange as a popular mode of Australian performance developed more fully in the PS practices of the 1980s and 1990s. Insofar as these activities have come to define an intercultural field of arts practice, they have also given rise to extensive debate within the PS community. Gilbert and Lo's critique of Australian intercultural performance in which they contrast collaborative and imperialistic tendencies is a case in point and one which underscores the argument we have been making. Imperialistic relations in performance, they argue, tend to focus on notions of 'Western cultural bankruptcy' on the one hand and specular, mystical otherness, on the other, while instances of collaboration embody a productive sense of tension (Gilbert and Lo, 2002: 39). As they argue, arts exchange between cultures is an important field of activity for PS, but the extent to which it has resulted in a suspension of cultural identity or even a process of questioning one's relationship to intercultural production is still to be determined (for example, see Eckersall, 2005; and Gilbert and Lo, 2002). Well-known pitfalls of intercultural performance include the tendency to reinscribe the fiction of identity in seemingly extreme and overt ways even when this is unintended. At the same time, we know that PS can develop more knowing collaborative practices that dissolve conventional orders and crack open the hermetically sealed world of national cultural essentialism.

A recent project exploring hybrid cultural tensions for PS was *Journey to Con-fusion* (1999–2003), a mixture of collaborative performance, workshops, seminars, and academic publication. *Journey* aimed to explore the relationship between aesthetic production and cultural politics in PS interactions within local-regional experiences of globalization. The project speaks to a progressive possibility within Australian PS to enact a form of 'ontological heterogeneity' (Félix Guattari, cited in McKenzie, 2004: 29). It represents one example of art performance practice and scholarship emerging in Australia which expressly counters (but does not always succeed in avoiding) the antipodal or 'imperialistic' model. As an extended project, *Journey* has had extensive analysis elsewhere (see Eckersall, Scheer, Varney, and Fensham, 2001; Eckersall, 2005; and collected authors in Eckersall, Uchino, and Moriyama, 2004). Here we will highlight perspectives arising from the project that contribute to the present discussion.

Journey to Con-fusion

Journey to Con-fusion developed from an artist workshop and critical seminar at the University of Melbourne in 1999. The Melbourne-based experimental performance group Not Yet It's Difficult (NYID) and

Gekidan Kaitaisha (Theatre of Deconstruction) from Tokyo worked for one week on skills exchange and performance showings. The companies met again in Tokyo in 2000 for workshops, a public showing, and a second conference. Creative development alongside intensive critical interactions shaped the outcomes of the work. Following the Tokyo season, Eckersall and Moriyama wrote about the project in upbeat terms, arguing that: 'The strategic realization and embodiment in performance of a double confusion/fusion way of thinking [...] lies at the heart of an alternative praxis in the present age' (2004: 21). In other words, over a time span of more than four years, the NYID-Kaitaisha project was a test case for a PS project in evolution: determined by experiences of collaboration and, as one participant said, a mix of 'indeterminable values' (Nishidô, 2004: 148). By the adoption of a combined critical-performative order, the project was specifically hoping to rethink experiences of antipodality in relation to global flows of culture.

In the volume exploring *Journey to Con-fusion*, Edward Scheer sees a PS model of virtual ecology evident in the work: 'A virtual ecology allows for the staging of the unforeseen if not the unpresentable, a horizon for all modes of becoming, performative or otherwise' (2004: 59). What matters, Scheer writes, 'is the production of ruptures' (2004: 60). As we argue here, ruptures and fissures that arise inevitably in cross-border collaborations are intrinsic to PS. They wind cultural and aesthetic threads of human experience together, not as finite gestures but as possibilities for mutation and change. Uchino Tadashi explores similar terrain in respect of the scholars working on the project and points to the importance of establishing divergent and diverse points of view. 'We should move fast enough so that there will be tons of different kinds of written records of such intracultural thoughts,' he writes (Uchino, 2004: 164). ' "Intracultural" not "intercultural" because if an age of globalization is an age of "Empire" in Hardt and Negri's sense (2000), there is no in-between space or clearly demarcated boundaries, but an all-encompassing intra-ness, to whatever cultural practice we are engaged in' (ibid.: 164). These arguments present two poles of an approach to PS where Scheer advocates a view in which alternative experimental practices might become active in the reshaping of culture more broadly in the region, while Uchino points to the limit conditions of any such cultural shift.

Journey to Con-fusion Part Three, shown in Melbourne in 2002, seemed to track both aspects of this approach. A work of intensity and violence – Vanessa Rowell wrote: 'By the end of the work there is an undeniable sense that the performers are trapped; hostages to the space, their bodies and their cultures' (2003) – it magnified small moments of action to become portentous. Denise Varney breaks into the evident symmetry of the gestic occupation of bodies in the performance as described by Rowell and moves into a reading of the work as 'Deleuzian becoming.' Her essay on the three performances of *Journey to Con-fusion* explores 'rhizomatic dramaturgy' as a basis for PS pointing to its amalgamation of aesthetics/dramaturgy/form and socio-political content. Varney reads the work as offering various kinds of escape:

> The performances are [...] a form of escape from contemporary Australia and Japan and their respective and collective histories and in this sense they evoke the Deleuzian becoming. By the second stage of the project, there is the sense of a temporary escape from Australia by the academics who journey to Tokyo [...]. By stage three in 2002, refugees seeking asylum have been detained in detention centres in Australia. The throwing of the victim against the wall sequence [seen in the performance ...] is both a reference to the violence in the detention centres and a mode of escape through performance from its empirical truth. [...] Each becoming is a reference to and an escape from cognitive truths through the symbolic. [...] It is a line of escape, 'an intense line of flight' to a place where the subject is no longer interpellated by his father/employer/state.
>
> (2004: 121)

This is not escapism but flight, a break out. Varney's reading of *Journey to Con-fusion* extends the potentials and simultaneously disorients PS thinking in ways productive to thinking about PS overall. The images in the performances described here represent experiences of occupation and systems of control that are enacted on the body. They also resonate as the after-effects of the constructed image and its meanings in the world. They extend the notion of performance and its relations to audiences in ways that emphasize PS as a kind of reverberating stammer. This is not to reprise a systemic critique of representation, although that is a part of Varney's work. Rather, to return to Scheer's terminology, it theorizes the performance event as an ecology of production. Varney's identification of the PS double as a 'reference to/escape from' restates the core problematic here, but suggests that art and culture are not a closed system and instead offer something less programmed, more potentially and usefully disruptive.

Conclusion: the here and the elsewhere

One of the aims of *Journey to Con-fusion* was to 'remap cultural globalisms from the south' (see Tsoutas, 2005). Increasingly, even while such types of cultural development remain in some way an inchoate endeavor,[3] it is projects such as *Journey* that indicate the possibilities for PS in Australia in terms of non-antipodal and non-imperialistic modes of cultural engagement. But there is a profound sense of incompleteness to the PS project of cultural engagement as it has evolved thus far, an absence which future study in PS will have to negotiate. In the context of the broad failure thus far to engage meaningfully with indigenous performance forms there are a number of significant exceptions: recent work by Maryrose Casey in aboriginal theatre (2004); Stephen Muecke's ongoing analyses of indigenous forms of thinking and narrative (2004b; and also see Muecke with Shoemaker, 2001) which, while not instantly recognizable as PS, perform some similar tasks; and Inga Clendinnen's *Dancing with Strangers* (2003) addresses the potentials of this relational scholarship in terms of a whimsical history of the colony.

In this sense PS in Australia is the story of a somewhat immature and undisciplined discipline, one that is not an exact reflection of the origins of the field in the United States, but which maintains an active reinterpretation and interconnection with a diverse set of theories and methodologies. Thoughts of performance anxiety ('are we doing it right?') are counterpoised by the very real pleasures of knowing that there remains enormous scope for possibilities for new kinds of performative and intellectual work, and in the freedom that arises from the fact that, from the point of view of the United States and Europe, PS in Australia remains both over the horizon and under the radar.

So the antipodean story may still serve as an explanation, however partial, of the difficulty of determining 'whether there is such a thing as "Australian Performance Studies,"' and 'if so, how it differs from Performance Studies in other geographical locations, and especially from the United States.' Antipodality is not contingent upon the circumstances of the Performance Studies international (PSi) conferences, but is historically determined by much larger and deeper factors. It is also a trap. The only way forward, as Morris says, is to move beyond 'antipodean inversion games' and to embrace a suitably performative solution, 'the art of improvisation' (Morris, 1984: 7). Improvising culture and identity in PS and more generally is to focus on interrelations, intersubjectivities, and intercultures. It is to lose the sense of grumpy self identity or reluctant antipodality and to cross territories of thought without an eye on the

map, to get lost in the 'permanent ironic play of similarity and difference, the familiar and the strange, the here and the elsewhere' (Clifford,
1988: 146). Sublime con-fusion may be the first appearance of the new.

Notes

1. There are a number of named performance studies programs in Australian
   universities and other research and teaching happens in the context of wider
   programs in arts, humanities and creative arts. For example, the UNSW, Victoria University, Monash University, and Sydney University identify that they
   have teaching programs entitled as 'performance studies,' while other large
   programs at Melbourne University, Deakin University, and Queensland University include performance studies under the rubric of theatre studies and/or
   drama. In reality, most programs of performance studies are interdisciplinary
   in nature, and it is perhaps Victoria University and Sydney University that have
   the most focused performance studies programs, although they each possess
   different intellectual compositions. No one model can stand for PS in Australia.
2. Three research journals cover the field of performance studies, each with
   differing flavors. *Australasian Drama Studies* is the oldest extant theatre and
   performance journal and was established in 1982. It features research articles,
   interviews, and performance documentation. While not specifically orientated
   to performance studies, *ADS* was the only referred source of specialist publication until recently. *About Performance* is a specialist performance studies journal
   published by the Department of Performance Studies at Sydney University. *Performance Paradigm*, a journal of performance and contemporary culture and
   co-edited by Edward Scheer and Peter Eckersall, was established in 2005. It
   aims to give space to critical perspectives in national, regional and global locations of discourse. The editorial board draws as much on Asia-based scholars
   as those in Western traditions, and the journal aims to publish a diversity of
   scholarship.
3. See the essays in Tsoutas, 2005.

References

Benjamin, W. 'Surrealism: The Last Snapshot of the European Intelligentsia,' *New
  Left Review*, 108 (1978) 47–56.
Casey, M. *Creating Frames: Contemporary Indigenous Theatre 1967–97* (Brisbane:
  University of Queensland Press, 2004).
Cioran, E. *The Temptation to Exist*, trans. R. Howard (London: Quartet, 1987).
Clendinnen, I. *Dancing with Strangers* (Melbourne: Text Publishing, 2003).
Clifford, J. 'On Ethnographic Surrealism,' in *The Predicament of Culture: Twentieth-
  Century Ethnography, Literature, and Art* (Cambridge, MA: Harvard University
  Press, 1988), pp. 117–51.
Dyer, R. *White* (London and New York: Routledge 1997).
Eckersall, P. 'Theatrical Collaboration in the Age of Globalization: The Geki-
  dan Kaitaisha-NYID Intercultural Collaboration Project,' in *Diasporas and*

Interculturalism in Asian Performing Arts, ed. H.-k. Um (London and New York: RoutledgeCurzon, 2005), pp. 204–20.

Eckersall, P. and N. Moriyama. 'Introduction,' in *Alternatives: Debating Theatre Culture in an Age of Confusion*, ed. P. Eckersall, T. Uchino, and N. Moriyama (Brussels: PIE Lang, 2004), pp. 7–22.

Eckersall, P., T. Uchino, and N. Moriyama. (eds) *Alternatives: Debating Theatre Culture in an Age of Confusion* (Brussels: PIE Lang, 2004).

Eckersall, P., E. Scheer, D. Varney, and R. Fensham. 'Tokyo Diary,' *Performance Research*, 6:1 (2001) 71–86.

Forbes, J. *Damaged Glamour* (Sydney: Brandl & Schlesinger, 1998).

Gilbert, H. and J. Lo. 'Towards a Topography of Cross-Cultural Theatre Praxis,' *The Drama Review (TDR)*, 46:3 (2002) 31–53.

Hage, G. *White Nation: Fantasies of White Supremacy in a Multicultural Society* (Sydney: Pluto Australia Press, 1998).

Hardt, M. and A. Negri. *Empire* (Cambridge, MA: Harvard University Press, 2000).

McKenzie, J. 'The Liminal-Norm,' in *The Performance Studies Reader*, ed. H. Bial (London and New York: Routledge, 2004), pp. 26–31.

Maxwell, I. 'Parallel Evolution: Performance Studies at the University of Sydney,' *The Drama Review (TDR)*, 50:1 (2006) 33–45.

Morris, M. 'Two Types of Photography Criticism Located in Relation to Lynn Silverman's Series,' *Art & Text*, 6 (1982) 61–73.

———. 'Des Epaves/Jetsam,' *L'Australie: d'un autre continent*, exhibition catalogue (Paris: Galérie de l'Arc, 1983). Reprinted in *On the Beach*, 3:4 (1984) 2–7.

Muecke, S. *Ancient and Modern: Time, Culture and Indigenous Philosophy* (Sydney: UNSW Press, 2004a).

———. *Aboriginal Australians: First Nations of an Ancient Continent* (London: Thames & Hudson, 2004b).

Muecke, S. with A. Shoemaker. *Legendary Tales of the Australian Aborigines: David Unaipon* (Melbourne: University of Melbourne Press at the Miegunyah Press, 2001).

Nishidô, K. 'The Journey to Con-fusion: Between Australia and Japan,' trans. M. Eglington Satô, in *Alternatives: Debating Theatre Culture in an Age of Confusion*, ed. P. Eckersall, T. Uchino, and N. Moriyama (Brussels: PIE Lang, 2004), pp. 143–8.

Plato. *The Dialogues of Plato, Vol. 3: The Timaeus and Other Dialogues*, trans. B. Jowett, ed. R. M. Hare and D. A. Russell (London: Sphere Books, 1970).

Rowell, V. 'Untamed and Trapped' (review of *Journey to Con-fusion #3*), *Realtime*, at: http://www.realtimearts.net/nextwave/rowell_confusion.html 2003 (accessed 30 July 2006).

Scheer, E. 'Dissident Vectors: Surrealist Ethnography and Ecological Performance,' in *Alternatives: Debating Theatre Culture in an Age of Confusion*, ed. P. Eckersall, T. Uchino, and N. Moriyama (Brussels: PIE Lang, 2004), pp. 55–62.

Tsoutas, N. (ed.) *Knowledge+Dialogue+Exchange: Remapping Cultural Globalisms from the South* (Sydney: Artspace, 2005).

Uchino, T. 'After 9.11,' *Alternatives: Debating Theatre Culture in an Age of Confusion*, ed. P. Eckersall, T. Uchino, and N. Moriyama (Brussels: PIE Lang, 2004), pp. 163–6.

Varney, D. 'Rhizomatic Dramaturgy,' in *Alternatives: Debating Theatre Culture in an Age of Confusion*, ed. P. Eckersall, T. Uchino, and N. Moriyama (Brussels: PIE Lang, 2004), pp. 117–26.

7
Critical Writing and Performance Studies: The Case of the Slovenian Journal *Maska*

Bojana Kunst

It is merely a coincidence that I am writing the present text at a time when the Slovenia-based magazine for contemporary performance, *Maska*, is preparing to celebrate its 100th issue.[1] Bearing in mind that *Maska* attempts to theorize contemporary art practices, this jubilee may seem a little suspicious, especially if we read it as a confirmation of a certain temporal continuity. But the journal's editorial board understands the publication of the hundredth issue and the accompanying celebrations principally as an opportunity for staging a strategic gesture of visibility that has little to do with continuity, but everything to do with developing a parallel theoretic and artistic practice. Behind this jubilee thus lie interesting narratives about the strategic ways in which a field of theorizing and of practicing contemporary art is constituted. The first will be a story about mounting a different platform for critical reflection on art and culture, where the journal not only represents certain artistic practices but also participates actively in them. The second is a story about the relation between the 'local' and the 'global,' which needs to be complemented here with the category of the 'national,' due to the specific post-socialist and transitional circumstances of Slovenia. The context of *Maska* should also be considered in relation to similar initiatives in the area of what was Yugoslavia, the country of which Slovenia was once a part.[2] The third will tell of the construction of a place for knowledge that is located outside academic institutions; and although it has adopted certain characteristics of academic practice, Maska has always been in close connection with the live practice of art.

This somewhat schematic breakdown of what have been a complex and dynamic two decades of critical thinking and writing will not only help us to gain insight into the ways in which a certain mode of thinking about art is constituted, but also into how a certain 'discipline' of thinking is established. Despite the fact that it is unceasingly committed to the dynamic practice of art and the practicing of resistance (as Foucault understands it, in the sense of a primary force; see Foucault, 1990), it does not forego rigor. In this instance we do not speak of an academic discipline of thought (as I understand performance studies to be), but rather of a material practice of thinking, which is nevertheless still closely connected to certain aspects that characterize performance studies. Such is the perspective from which we today, on the eve of publishing the hundredth issue of *Maska*, look back at its history.

Singularizing the name and the apparent continuity of time: a brief history of *Maska*

The hundredth issue of *Maska* (published in September 2006) is conceived of as a publication of projects and their critical reflection as if they were taking place in the year 2023 (when the 200th issue of *Maska* would be published). The issue includes artists and scholars who have written for *Maska* or about whom *Maska* has written over the past 15 years. This event is not by coincidence imbued with a utopian sense of futurity. *Maska's* hundredth issue will be published as the 200th and will simultaneously celebrate the two jubilees, both of which are, in fact, entirely imaginary. They may be understood as resulting from a certain strategic gesture of visibility enacted by a specific art practice and its way of thinking. This gesture is not, though, to be mistaken for a vision of utopia as the historical avant-gardes introduced it, in which the present is completely submitted to the future. Rather, what is at stake is the consistency of the gesture with the very manner in which the editors, writers, and other collaborators have understood the role of *Maska* thus far.

To understand this gesture we have to outline briefly the history of the magazine. *Maska* published its first edition as far back as the year 1920, when the project was launched by a movement of theatre directors, playwrights, and actors holding so-called 'progressive' views, who wanted the magazine to revive the situation in Slovenian theatre. Under the influence of contemporary avant-garde currents in Europe, the first issue of *Maska* was published with the expressed intention to transform art and life, about which we may read in the introduction: 'We must build the new human race. All nationalities build. Let's work. Let's build.' (cited in Hrvatin, 2006: 492) Yet facing conservative opposition and, later, totalitarianism, the initiative quickly came to an end and was only revived in the 1980s. In 1985, the magazine *Maske* ('maske' is the plural of the Slovenian word for mask, 'maska') was established under the editorship of Peter Božič and Tone Peršak. It set out to advance reflection on current theatrical events and to establish a link between professional and amateur performers. As this was the time when many new theatre practices were being experimented with across numerous fields, these new formats quickly found their way into the open and pluralistic nature of the magazine.

Parallel to this, another ambitious publication project was being formed. *Euromaske* (under the leadership of theatre director Dušan Jovanović, theatre historian Dragan Klaić, and editor Peter Božič), with a rich visual style and English-language content, internationalized the magazine but only brought out three issues.[3] In this, *Euromaske* preceded other, similarly unsuccessful attempts (for example, *Hybrid Magazine,*[4] *TheaterSchrift,*[5] or the English version of *Ballet International/Tanz Aktuell*[6]) at publishing in a more and more connected European space, where an increasingly global cultural industry paradoxically occasions less and less room for common initiatives of thought. In the early 1990s an important shift in the editorial policies of *Maske* took place, when the magazine was taken over by the new editorship of Irena Štaudohar and Maja Breznik (then students of dramaturgy at the Ljubljana Academy of Theatre). An editorial board was formed of artists and scholars whose initiatives were prevalent in the Slovenian contemporary performance scene at the time. *Maska*, as we know it today, is a result of this shift and has been published since 1998 under the editorship of Emil Hrvatin (in 2002 it became bilingual and now publishes in Slovenian and English) and will, following the hundredth issue, from 2007 on be edited by a member of the youngest generation of editors, Katja Praznik.

This brief overview serves not merely to provide factual faithfulness to a globally marginal and invisible history. I am even less concerned with the construction of a sense of historical continuity. Quite the contrary: the platform for writing and for knowledge production that *Maska* offers today is principally the result of a history of breaks and of resulting interpretations of its own fictional continuity. We are dealing with the story of the singularization that took place at the beginning of the 1990s when magazine *Maske* (plural) became *Maska* (singular) again.[7] The journal adopted the name from the remote period of the historical avant-garde and began to label itself as its continuation, yet at the same time furnished itself with a new motto: 'Theatre is the first to unveil the time to come.' Repetition in time is here not a reference to a particular progressive gesture in the past, but rather points to the fact that we are dealing with an act of singularization. This can be understood as a break, but in a manner which rearticulates the history of the break itself. It constitutes a platform which positions itself as new in a rearticulated historical context. The break itself, the very gesture of the new, is therefore already being historicized. *Maska* systematically affirms itself as continuity, although this continuity is, as a matter of fact, constructed as constituting a discourse on art that is different.

It is interesting that the groundbreaking art movements included in the collective NSK Neue Slowenische Kunst (i.e., Irwin, Scipion Nasice Sisters Theatre, New Collectivism, Laibach), which opened up Slovenian cultural space for new interdisciplinary, postmodernist practices of art, acted in a similar way during the 1980s and in the first half of the 1990s. Their activity was deeply marked by the principle of *retrogardism*, in which the deliberate return to the motifs of the historical avant-gardes and to the utopian affirmative potential of twentieth-century art is interwoven with postmodernist ways of thinking. In other words, such retrogardism links the enactment of a total work of art through the synthesis of its particularities (which can be found principally in the historical avant-gardes, in the concept of a total work of art, and in the interpretation of national myths) with the enactment of the particular and processual form of the artwork (which can be found mainly in postmodernist procedures of coincidence and the open-endedness of structure). This practice therefore had an ambivalent effect, archaic and contemporaneous at the same time, for it combined two different politics of an affirmative present. The first one is utopian, where the new will replace the old; the second one is the elusive formation of an aesthetic equality, where there is no submission to a single principle and the formation takes place through the immanence of the material itself. The connection between the two principles is thus neither the result of postmodernistic eclecticism, where 'anything goes,' nor the baroque restoration of the historic memory of socialist Europe (as many interpreters from Western Europe understood it initially). It is within this context that the singularization of the name of *Maska* magazine has to be understood. From 1990 on *Maska* becomes an important document of the new generation that emerged from this paradigmatic break. The first issue under the new editorship in the 1990s was dedicated to the groundbreaking performance *Baptism Under Triglav* (directed by Dragan Živadinov), made in 1986, considering it from a temporal distance whilst constituting a platform on which new artistic practices could be established.

The singularization of *Maska's* name therefore established continuity as a strategic gesture, with which it directed attention to the important history of its local context and, at the same time, gave way to new ways of thinking. As outgoing editor Emil Hrvatin writes:

> I have never viewed *Maska* as a document about a specific (artistic) practice but rather as a specific way of thinking. It is a document about the practice of thinking. Only as a practice of thinking it can become a document about an artistic practice. The practice of thinking is not something that would naturally follow artistic action in its discursive form. At *Maska*, the practice of thinking is inseparably linked to the practices of thinking about art. *Maska* has never been at the outside of artistic practice; in fact, together with art, it constantly rearticulates its position and territories, constantly inhabiting the space in-between, where relationships of power have not yet been established.
>
> (Hrvatin, 2006: 491)

The intimate link he draws between thinking about art and artistic practice sheds a particular light also on theory itself, which *Maska* has understood primarily as a material practice that at the same time discloses different discourses and the spaces and strategies of art. Such a role of theory, which inhabits spaces in-between, opens up performance as an interdisciplinary space of contemporary idioms, activities, and artistic and cultural procedures. It thereby approaches the interdisciplinary aspects that have characterized performance studies, especially in the second half of the 1990s. This role of theory also marks the internationalization of *Maska* that has occurred over the past few years (*Maska*, as we noted, being published from 2002 on as a bilingual edition), which opens the door to a thinking between institutional, academic, and practical initiatives.

A joint issue of *Maska*, *Frakcija*, and *Performance Research*, published in 2005, bears the title 'On Form/Yet to Come,' attempting to delineate a certain cultural and research dimension that issues from the intriguing dialogue between these publications. I have described the basic characteristics of this dialogue in an article written on that occasion:

> The important outcome is thus the acknowledgement that art production is tightly connected to institutionalization and commercialization, and that it succumbs to similar bureaucratic laws and participatory problems as exist everywhere. [...] The strategies are not directed towards the discontents of the law, but toward the very

ability of power and control, toward the subtle and overwhelming mechanisms by means of which power and control are entering our contemporary life and contemporary work.

(Kunst, 2005: 46)

But in order to understand *Maska* as a platform for the production of knowledge that is closely connected with certain aspects of performance studies, we should first clarify the context within which it is articulated. This context may be described as a specific relation between the local and the global, which opens up the space of collaboration as a space in-between.

Between the local and the global: a story about the politics of affection and uneasiness

Maska constituted itself at the beginning of the 1990s not only as a stage for a particular generation of artists who were developing new concepts of performance. It also generated visibility for this artistic practice both in the local and the global context. At the same time *Maska* introduced into the Slovenian cultural space new theoretical approaches and expanded the field of thinking about contemporary art with translations of texts and documents, with guest lecturers, and lately also with translations of important books.[8] It is necessary to mention at this point that from the beginning of the 1990s until today, Slovenia has not developed any systematic academic disciplines dedicated to the study of contemporary art and culture.[9] With its internationalization in the year 2000 (and with some of the bilingual book editions), *Maska* also entered the European cultural space as a magazine that opens up new critical platforms of thinking about art. But this development is not unproblematic. The concomitant problems may provide an interesting analysis of the multiplicity of relations that exist between the local and the global, particularly as we attempt to reflect on performance studies as a kind of 'global phenomenon' of the 1990s.

   Maska emerged from a specific local political context, namely the post-socialist and transitional cultural milieu of Slovenia, which has been marked by proximity to the tragic war in neighboring Croatia and Bosnia-Herzegovina. In contrast to other parts of former-Yugoslavia, Slovenia drifted into war for a very short period only. An entire generation of artists and thinkers was connected with this transition – but not in the role of opposition, as many interpreters from Western Europe incorrectly concluded. Ex-Yugoslavia had always been an incomparable phenomenon, with a socialist system combining the rigidity of the East with elements of the consumerist hedonism of the West. In as early as the 1960s, Yugoslavia produced a highly developed modernist art practice. The new generation of artists in the 1980s was not interested in oppositional approaches to art, nor in the creation of a politically dissident message. It developed its aesthetics with the help of highly affirmative formal artistic procedures, together with a reconceptualization of art and its contexts. Something similar was taking place in the field of theory, where the so-called Ljubljana Psychoanalytic School shifted the field of philosophy into an interdisciplinary study of theory, cultural phenomena, philosophy, and art.[10] The processes of art and culture were marked by interdisciplinarity, formal variety, and reconceptualization. In these processes we may find a certain specificity, which I would describe as a continuous awareness of the material practice of art. The art movements of the 1990s thus had a profound effect on the way in which the dynamic field between reflection, study, and critical writing developed. Hereafter, the psychoanalytical movement turned to more traditional forms of art and to film, while *Maska* developed writing on a broader notion of contemporary performance. But both initiatives attached great importance to the cultural-political contexts of art.

The particular relation between the local and the global in which *Maska* attempts to act and think reflects critically on specific artistic histories of the twentieth century, which are invisible both in the local and in the global context. *Maska* has always tried to connect the local with the recognition of particular cultural and artistic initiatives, to bring into view how this space possesses a strong 'history of contemporaneity.' This term is deliberately contradictory, while it delineates the search for, and the recognition of, art initiatives that developed in close connection with the contexts from which they emerged. We cannot place them in the linear history of modernism, marked by aesthetic and emancipatory progress, which in its verticality distinguishes between visible and invisible geopolitical territories of art.[11] Rather, what is at stake here are different modes of activity, where art is at the center of social, cultural, and political processes. As one of the active members of the editorial board of *Maska*, I have always understood this historicizing as a positioning in the space in-between, in the conflicting intersections between many cultural contexts and histories. Their recognition enriches our material practice and opens it up to formal varieties, constitutes a horizontal field of connections between similar initiatives, or, in other words, functions as a unique critical network, through which the mutual spaces of a different public may be disclosed.

Activity of this kind intentionally differed from the processes that in the majority of cases marked the post-socialist and transitional cultures in the countries of Eastern Europe. In *Postmodernism and the Postsocialist Condition*, Aleš Erjavec describes these processes as follows: 'with the transition out of post-socialism and toward capitalism, the majority of the artists [...] were forced to adapt to the emergent world of capitalist aspirations in their own countries or be marginalized. In this respect the position of these artists was, of course, not much different from that of other intellectuals of the period or of ordinary people who were also forced to reinvent their lives' (Erjavec, 2003: 29). In the circumstances of that transition, artists supposedly favored globalized cultural artifacts over subversive and politically provocative art. I would like to mention here the outrightly absurd title of a festival in Rotterdam in 2005, which presented young artists from the countries newly associated with the European Union under the name 'Paradise Regained?,' to help understand how the narrative of transition and reinvention prevails in relation to the territory of Eastern Europe. There is a problematic conception of a vertical linear history at work here, which presumes that countries in transition are abandoning their socialist past (in which capitalism is yet-to-come) to reach the final goal that will place them on the contemporary map. The process of transition thus makes us all the same in the contemporary global paradise of redistribution and connection.

Maska, however, has always tried to think the processes of art through more complex connections and has problematized the mirrors that we hold in front of each other during these processes. One of the basic problems is the way in which we may think connections and networks among different artistic initiatives as an affirmative process which creates a different public. In this process, reinvention is not equated with appropriating models of success on the global cultural market. It results instead from the articulation of different energies and the creation of events which are not necessarily of our common temporality. The interdisciplinarity of art events in the 1990s was understood also in close connection with an in-betweenness, where the Other no longer understands itself as a déjà-vu, as a belated, obsolete linguistic being, but constitutes itself within shifting individual and collective differences. This in-betweenness enables different practices to be articulated, at once inside and outside, and become manifest as new artistic activities. We can say that today contemporary critical writing, as well as contemporary performance, takes place mainly in such spaces in-between. Here a continual negotiation between national experience, community interests, cultural values, and particular histories can be detected. Performance thus occurs in the cross-section between local and international space but also in the cross-section between particularity and universality.

With this kind of positioning *Maska* is comparable to other theory initiatives in former Yugoslavia, such as the journal for performing arts *Frakcija* from Zagreb and the magazine *TKH (Walking Theory)* from Belgrade. *Maska* is closely connected with them through numerous projects: conferences, research platforms, educational projects, international collaborations, publishing of joint issues, and so on. These projects have opened up a different notion of locality, which is no longer understood as a topographic inscription between the developed European West and the undeveloped cultural territory of the East. One of the main problems in understanding art originating from Eastern Europe is, as Boris Groys points out, that 'Eastern European always comes from Eastern Europe, it is always seen as information on the state of the society of its origin' (Groys, cited in Pristaš, 2005: 8). This political representability (where one always performs in the role of the Other) has nothing to do with the described activities in the spaces in-between, with the creation of events, with discursive self-legitimization. The platforms of thinking, writing, and creating which developed around *Maska* and its partner magazines understand the potentiality of globality neither as a possibility for reinvention or multicultural sympathetic recognition, nor as a process of transformation and conformity to the global art and cultural market. In my view, this is the only way to think global connections – namely, as a network of connections which renders possible the practicing of 'how to be outside.' 'The act of discursive self-legitimization is not the act of a solution to go beyond boundaries but to legitimate the outside as a strategy both of appearance and disappearance. The act of going and the act of staying' (ibid.: 8).

A platform for the production of knowledge

Maska also directly connects with certain aspects of performance studies, in terms of methodology as well as of content. However, *Maska* has never understood the study of performance as an academic practice, but as a material practice, a practice of research into different aspects of performing. As I attempted to show in the previous section, due to its specific local context, the critical writing of *Maska* had to face rather abruptly the processes of geopolitical performance and to re-examine its own position among the images that the East and the West of Europe were showing to one another at the beginning of the 1990s. Slavoj Žižek describes the symptomatic extremes of these images as follows:

> The disappointment was mutual: the West, which began by idoliz-ing the Eastern dissident movement as the reinvention of its own tired democracy, disappointedly dismisses the present post-socialist regimes as a mixture of corrupt ex-communist oligarchy and/or ethnic and religious fundamentalists. [...] The East, which began by idolizing the West as the model of affluent democracy, finds itself in the whirlpool of ruthless commercialization and economic colonization.
>
> (Žižek, 1999: 40)

Slovenia was successful in avoiding both these extremes. This does not mean, however, that it did not face its own problematic. In the 1990s the processes of art and culture were articulated right amid the performed topography of the Other and the search for its apparent authenticity, which was slowly diluting under the 'equality' of globalization. The recognition of these processes was furthered by performance studies, which *Maska* developed with the help of translations, conferences, and lectures. Performance studies offered an analysis of artistic and political phenomena through the theory of the performative, and an explo-ration of the processes of performativity and their political ramifications. On the one hand, this critical project succeeded in positioning artistic practices in the context of spaces in-between, on the other hand, the artistic practices of the 1990s themselves moved between genres and were actively present in the international scene.

At the end of the 1990s it became evident that the cultural and artistic atmosphere in Slovenia was drastically changing under the increasing pressure of commercialization and as a result of the unsuccessful local institutionalizations of certain initiatives for artistic production and edu-cation. As a matter of fact, in the Slovenian cultural as well as political space we may have witnessed the paradox of normalization, which despite its specificities is part of a more general problem that contempo-rary Western societies too are facing. Over the last few years this process of normalization – which has taken place under the banner of economic and political success, the introduction of democratic processes and global recognizability – has marginalized civil and cultural affirmation and ways of critical thinking and doing, whilst it has enacted an empty form of tol-erance. This process of normalization has also influenced the conception of artistic subjectivity itself, where, as Susan Buck-Morss proposes, 'the artistic freedom exists in proportion with the artistic irrelevance' (2003: 69). In other words, it seems that today, artistic and creative powers are isolated from social efficacy by the very fact of their normalization. *Jouissance* of the private and the arbitrariness of everyday life seem to be the center of post-capitalist production. The normalization of artistic subjectivity is revealing the exhaustion of subversive and transgressive modes, which have become an intrinsic part of contemporary commodification. The commodified *jouissance* of the private has unfortunately lost its potential for revolt in the commodified *jouissance* of global happiness. The real question at stake is therefore not any longer how to find a way out, but how to develop procedures of disobedience to the process of normalization.

The cultural and social processes of the last few years have indeed revealed the necessity to rethink the material practices of theory and of art. And this necessity has deeply marked *Maska's* activities. With educational projects, book publishing, the production of artistic events and international collaborations, *Maska* has expanded its activities from a magazine into a wider platform, which I understand to be an important locale for the production of knowledge and critical thinking. This step has also advanced *Maska's* internationalization, for the magazine finds itself facing the need to examine and reflect on the changes in these material practices that have taken place with the processes of globalization: democracy becoming a spectacular way to maintain global disorder, the exhaustion of democratic ideas, the collapse of multiculturalism, and the disappearance of the critical public. The place of artistic labor is nowadays increasingly located in the center of post-industrial processes of immaterial production. This kind of normalization is linked to another, entirely global characteristic: an understanding of life that is being progressively appropriated and regulated by global capital and economy. Today, life itself is located in the center of the materiality of economic processes. The aesthetic competences of intensities, energies, and events are at the heart of the contemporary commodification of entertainment.

Critical thought thus needs to rediscover a way in which to articulate the potentiality of ways of life, which may bring about a change in the ontological place of art itself. It is necessary to examine both the *poesis* as well as the *praxis* of art. Is it still possible to think artistic practice in the context of the current appropriation of life as a unique potentiality? Paolo Virno (2004), theorists and activist of the Italian movement *Autonomia* and one of the first analysts of the performative traits of post-Fordist working processes, has underscored some of the characteristics implied in this shift. Among them is the fusion of labor and political action, where contemporary labor has adopted many characteristics that used to pertain to political practice. *'Poesis* has taken on numerous aspects of *praxis*,' states Virno (2004: 50), by which he means that labor has adopted the previous characteristics of public activity, such as virtuosity and performing, as well as many characteristics of artistic activity proper. Labor has become an activity without a final goal, yet which unceasingly demands the presence of the Other. Therefore it makes sense only insofar as this activity can be heard and seen. Labor today is closely connected with cognitive qualities and linguistic competences, and with the ways in which labor is understood in art practices. We can thus say that the field of performance studies, at least in the case of *Maska*, is now recontextualized as a mode of thinking about the links between action, labor, and the political, and their close connection with practices of thinking and writing about art.

Conclusion

Performance in its broadest sense has found itself during the last few decades at the center of various debates in art, history, and society, including politics, science, and technology. At the same time, performance has become the focal point for understanding locality and internationalization, which is, at least in the instance of Slovenia, closely connected with questions of space and the institutionalization of individual initiatives. To put it briefly, performance in the broader sense is closely connected with the comprehension of space, the context of its emergence, and the demand for visibility in the local and global context. The critical writing in *Maska* has positioned itself in direct connection with performance as a political rearticulation of thinking and practicing art. It is necessary to consider another aspect of performance here, which relates primarily to the potentiality for the real that may be articulated with this kind of performing. The editor of the Belgrade magazine TKH (*Walking Theory*), Ana Vujanović, has described *Maska's* activities as a sort of 'hacking' of the virtual, whereby she understands 'hacking' as an opening up of closed zones, with the intention to transform their procedures or protocols, and 'the virtual' as the unrealized potentiality of the real. 'During the past fifteen years *Maska* did not, in all these incomparable contexts, represent the Slovenian performing arts scene and their progressive practices, as it did not represent a new model of discourse production, not even a critical one. What it pursued was the continual "hacking of the virtual" ' (Vujanović, 2006: 86). She draws attention to the inevitable contingency of this kind of practice and to the multiplicity of its strategies, which position *Maska* in the unstable processes of persistent rearticulation, in the continual entering into and withdrawing from the performance of different material practices. Paradoxically, it is precisely such instability that allows for 'the opening of the current state of affairs to its potentialities' (ibid.: 86). For *Maska*'s 'hacking' is a perpetual multiplication of the relation between the actual and the virtual within each of the stated contexts and their potentialities.

Herein I find the power of critical writing today, and this is also how I understand the invitation to write a contribution for a compendium that finds its common denominator under the title 'Contesting Performance.' To participate in a certain field of knowledge makes sense only when the knowledge in question opens the potential for a certain unrealized thought of the Real. In this sense *Maska*'s relation to the research of performance can be understood as one of the numerous contexts in which such a potential may emerge. At the same time, to open the potentiality of a different temporality points to the fact that we are no longer dealing with the utopian transformation of life through art, which was the characteristic of the twentieth century. Rather, we may speak of articulating possibilities for the actualities in which we live, of the struggle for imagination, language, and intensities, and of revealing a role for contemporary performance that may enable a common enactment of the Real.

Notes

1. *Maska*'s hundredth issue was published in September 2006 under the title *Do-it-yourself*.
2. *Maska* is the only regularly published magazine for contemporary performing arts in Slovenia. It has been published as a bilingual publication in Slovenian and English since 2002. Slovenia is one of the younger European states. It was a part of former Yugoslavia until 1991, when it attained independence, and it became a full member of the European Union in 2004.
3. *Euromaske* no. 1 was published in the fall of 1990, no. 2 in winter 1990, and no. 3 in spring 1991.
4. *Hybrid Magazine* was a performance magazine published in the UK, which ran between 1992 and 1994.
5. *Theaterschrift* was published between 1993 and 1998 by a consortium of European performance spaces. It appeared in four languages (Dutch, French, English, and German).
6. A separate English edition of *Ballett international/Tanz aktuell* was published between 1994 and 2001. The magazine, renamed *BallettTanz*, is now published in a bilingual edition.
7. *Maska* means 'mask' in English.

8. Two collections on the *Theory of Contemporary Theatre* (1996) and *Theory of Contemporary Dance* (2001) were published under the editorship of Emil Hrvatin. Among other book projects are the Slovenian translation of H. T. Lehmann's *Postdramatic Theatre* (2003) and Amelia Jones's *Body Art: Performing the Body* (2003), studies on contemporary Slovenian art (by Inke Arns, Alexei Monroe, and Tomaž Toporišič), and bilingual books on artists such as Davide Grassi and Emil Hrvatin.

9. Slovenia does not know the academic discipline of performance studies, nor the German model of *Theaterwissenschaft*. The Ljubljana Academy of Theatre offers a program in 'Dramaturgy,' but the studies are bound to a traditional conception of theatre and to an understanding of the role of the dramaturg as the analyst of the dramatic text and the contexts of its performance.

10. Today this movement is known principally through the philosopher Slavoj Žižek. Some of its other representatives are Mladen Dolar, Miran Božovič, Alenka Zupančič, and Renata Salecl.

11. Thus it has become somewhat of a truism that the former socialist territory of the European East does not have a developed contemporary art scene as it did not go through the experience of modernism.

References

Buck-Morss, S. *Thinking Past Terror* (London and New York: Verso, 2003).

Erjavec, A., 'Introduction,' in *Postmodernism and the Postsocialist Condition: Politicized Art Under Late Socialism*, ed. A. Erjavec (Berkeley: University of California Press, 2003), pp. 1–55.

Foucault, M., *The Care of the Self, History of Sexuality 3* (New York: Penguin, 1990).

Hrvatin, E. 'If there Is No Mission, a Mission Becomes Possible: On the Position of a Cultural Journal under Conditions of an Unbearable Lightness of Freedom,' in *Leap into the City – Chişinău, Sofia, Pristina, Sarajevo, Warsaw, Zagreb, Ljubljana: Cultural Positions, Political Conditions. Seven Scenes from Europe*, ed. K. Klingan and I. Kappert (Cologne: Du Mont Literatur und KunstVerlag, 2006), pp. 490–501.

Kunst, B. 'Yet To Come: Discontents of the Common History', *Performance Research*, 10:2 (2005) (joint issue with *Maska/Frakcija*, 'On Form/Yet to Come') 38–46.

Pristaš, S. G., 'Why do we produce ourselves, promote ourselves, distribute ourselves and explain ourselves? Why are we 'as well' around?', *Maska*, 20 (2005) 5–6.

Virno, P. *A Grammar of The Multitude* (Los Angeles and New York: Semiotext(e) Foreign Agents Series, 2004).

Vujanovič, A. 'Maskino hekiranje virtualnega', in *Sodobne scenske umetnosti*, ed. B. Kunst and P. Pogorevc (Ljubljana: Maska, 2006), pp. 78–87; here translated by A. Rekar.

Žižek, S. 'The specter is still roaming around – An Introduction to the 150th Anniversary Edition of the Communist Manifesto', *Frakcija*, 14 (1999) 40–1.

8
'Say as I Do': Performance Research in Singapore

Ray Langenbach and Paul Rae

Indonesia must first recognise that Singapore, as proclaimed by myself on behalf of the people and Government of Singapore today is an independent, sovereign nation with a will and a capacity of its own. A strong will, if I may be [*sic*] little immodest about the pride and the stamina of the citizens of Singapore in a capacity which is rather limited.

(Lee, 1965: 4)

So stated Prime Minister Lee Kuan Yew on 9 August 1965, in a televised response to queries from journalists about the regional implications of Singapore's ejection from Malaysia two years after full independence from the British. In a now iconic scene from the history of the Southeast Asian city-state, Lee went on to describe the separation as 'a moment of anguish' (ibid.: 21), before sensationally illustrating the sentiment by breaking down in tears: 'Recording was stopped,' states the official transcript delicately, 'for the Prime Minister to regain his composure' (ibid.: 22).

If there is little that is unique about a nation forged of performatives ('I proclaim [...]') and spectacles of anguish, Lee's press conference nevertheless hints at the distinctive context within which Singapore performance operates, and to which its analysis contributes. Formerly an outpost of the Riau-Johor Sultanate (fourteenth to nineteenth centuries), a British colony (1819–1943, 1945–63), and lynchpin in Japan's 'Greater East Asia Co-Prosperity Sphere' (1943–45), Lee's description of the island city-state as a 'nation with a will and capacity of its own,' on the day of its unforeseen inception, exemplified a stridently utopian approach to nation-building that has held sway ever since. Under continuous (and continuing) rule by Lee and his People's Action Party (PAP), Singapore's variegated pre-independence lineage of ownership, status, and allegiance would be selectively subsumed into a national teleology, whereby the state has continued to produce performative declarations and cultural performances around its own governmental performance of rapid economic growth against the odds. As such, any academic survey that, like the present volume, takes the nation-state as its geopolitical frame of reference must, in the case of Singapore, signal the degree to which it has been ideologically singularized as the pre-eminent concern of a pre-eminent political entity – the PAP.

Singapore as context: national, cultural, and administrative performance

Lee Kuan Yew had reason for anxiety in 1965. A densely-packed island with few natural resources and a multiethnic, multi-faith immigrant population located in a politically volatile region,[1] 'Singapore,' more than most postcolonial states, would need to be made up as it went along. Over many early press conferences, Lee literally talked a nation into existence on the international stage – a diplomatic and public relations exercise that continues to this day. Domestically, a capable, driven group of men masterminded Singapore's substantial economic fortunes and infrastructural development, and, concurrently, a national ideology developed that was both disciplinary and productive. A comprehensive program of legislation, surveillance, and propaganda ensured a compliant workforce and homogenized behavior, for purposes of national stability and economic survival. Myriad aspects of social life were engineered: personal hygiene and procreation habits, interpersonal and interethnic relations, language, the regulation of public and private space, and low-level militarization through a Swiss-style 'Total Defense' program. This was underwritten by a rhetoric of pragmatism that privileged actions over words, and brooked no objections on grounds of mere principle. In short, modern Singapore and the modern Singaporean have been, to a significant extent, performatively produced.

Predictably, this pragmatic performativity influenced what constitutes 'cultural performance' in Singapore. Historically, there was a strong philistine streak to Singaporean nation-building (in 1969, Lee famously stated that 'poetry is a luxury we cannot afford'). The theatre was seen as a vehicle of left-wing propaganda on the one hand, and so-called 'Yellow Culture' (decadent Western values) on the other. Today, excepting certain traditional forms (such as Chinese opera), all arts events require a license, and theatre companies, unless specifically exempted, must submit their scripts to the Ministry for Information, Communications, and the Arts

(MICA), where it is not uncommon for cuts to be stipulated to material with sexual, 'racial,' religious, or political content. Anecdotal evidence suggests that many instances of performance censorship go unreported, but a number of key events have been discussed extensively, and throw certain contours of the Singapore state into sharp relief. These include: the detention of leftwing theatre-makers in the 1970s (Wee and Lee, 2003: 18–19); the detention in 1987 of members of The Third Stage for their alleged involvement in a 'Marxist conspiracy' (Oon, 2001: 112); the decade-long proscription, from 1994, of performance art and Forum Theatre (Krishnan et al., 1996; Krishnan, 1997; Langenbach, 1996, 2003); and the banning of the English-language production of Elangovan's play *Talaq* (2000), about marital rape within the Indian Muslim community (Seet, 2002: *Talaq* Documents, 2001).

In 2000, the arts were granted political absolution of sorts, when MICA's 'Renaissance City Report' placed them center-stage in its vision of a globalized knowledge economy capable of attracting 'foreign talent' and generating local 'cultural capital.' State support for the performing arts increased, international events showcasing Singapore culture were staged, and local religious festivals (such as the Hindu *Thaipusam* and Muslim *Hari Raya Adilfitri*) and other spectacles (like the Chinese mid-autumn Lantern Festival) became widely promoted by the Singapore Tourism Board.

Meanwhile, state agencies have sought to foster national affect by producing their own performances. Most spectacularly, Lee's 'moment of anguish' is both commemorated and salved annually by the National Day Parade, which combines mass-participation displays, military drills, video screenings, and fireworks. Presented before an audience of thousands, the event also receives blanket television coverage, which uses an off-screen voiceover to allegorize every act in the performance (Langenbach, 2006). Nothing escapes interpretation (except the voiceover itself), and this characteristic of Singapore state operations is of particular significance to the performance researcher. For, while architect Rem Koolhaas was correct to write that 1965 represented 'a showdown between *doing* and *thinking*, won hands down by doing' (Koolhaas, 1995: 1033), we might further note how much that *doing* was also a *saying*, or, more precisely, a *stating*: a declaration and continuous articulation of the state.

Both described *and* enacted, the television spectacle of Lee's 'moment of anguish' initiated a rhetorical and performative operation of national amalgamation that, while most explicitly manifested in state cultural production, also had important consequences for intellectual production. Singapore inherited from the British a ramified civil service and bureaucracy whose expansion under the post-independence government was accompanied by regulatory interventions into all aspects of civil society (Buchanan, 1972: 239–311; Chan, 1997 [1975]; Chua, 1995: 9–46, 79–123). This led to the formation of a system that recalls Gramsci's 'statolatry,' a specific stage of the process of conforming civil society to the economy, in which, as Vacca glosses it, 'the bureaucracy becomes commonly understood as the *whole* state' (1982: 53). As early as 1975, Chan Heng Chee – a political science professor who, since 1996, has been on secondment as Ambassador to the United States – identified in Singapore an 'administrative state,' where the 'meaningful political arena is shifting or has shifted to the bureaucracy,' resulting in a 'steady and systematic depoliticization of a politically active and aggressive citizenry' (1997 [1975]: 294). The alignment of the bureaucracy with the administration was accompanied by that of many intellectuals in the civil service, universities, schools, and think-tanks, and of journalists working in a politically constrained media environment. Of the many consequences of these developments, two bear noting here. First, neither the pursuit of knowledge as an end in itself, nor the ideal of academic freedom upon which it is conventionally predicated, has held much sway in official attitudes towards the universities. The term 'research' is today synonymous in the public mind with the product-oriented and intellectual property generating processes that are key to driving a so-called 'Knowledge-Based Economy.' Second, the resulting performance indicators, based on impressive capital growth and labor efficiency, contribute to a nation-building panegyric in which the performance of the administrative state is not only to be assessed, but is itself the act of assessment.

Three factors therefore combine to inform the multiple meanings implicit in our chapter title: first, 'Singapore' is performatively produced; second, cultural performances are variously proscribed, tolerated, and promoted; and third, the administrative state (including the academy) simultaneously determines, assesses, and represents 'performance indicators.' As such, 'say as I do' describes an operation of illiberal democracy that is substantially more effective, efficient, and durable than the more dictatorial 'do as I say.' The pragmatic approach to nation-building valorizes actions before words, and the bureaucracy describes those actions appropriately. 'Say as I do' is also the demand that performances and performers make of their interpreters. Where does cultural performance research fit into this apparently all-encompassing context? And how might it suggest a way out? Latter sections in the chapter address these questions in turn.

Three types of performance research

International publications and state performativity

The idea that 'Singapore' out-performs its artists was popularly accepted by Euro-American postmodernists in the 1990s (William Gibson famously called it 'Disneyland with the Death Penalty' (1993), and Rem Koolhaas a 'theatre of the *tabula rasa*' (1995)), and was most coherently articulated by the Singapore-based British academic John Phillips:

> If there is an essential Singaporean identity apart and distinct from any other, it is that there is manifestly no essential Singapore identity apart and distinct from any other. But this is not to be found expressed, at least not in any direct way, in the writings, the fiction and poetry, the plays or the cinema, the music and dance of Singaporeans. Singapore is rather its own performance. No 'text' from Singapore should be regarded as simply representing Singapore's urban space, for urban space itself in Singapore's chief mode of representation.
>
> (Phillips, 2000: 188)

A literature professor, Phillips's textualization of Singapore is understandable, if contestable. More interesting is that a similar discursive occlusion of aesthetic practice is evident in many internationally written and published analyses of local performance. In *Theater and the Politics of Culture in Contemporary Singapore*, for instance, the American William Peterson is at pains to distance himself from the 'minutiae' of 'the relatively small field of theater in a small island nation' to look instead 'at a particular developmental model through the lens of culture and, more specifically, through theater' (2001: xi). Peterson systematically demonstrates how theatrical practice and political discourse were intertwined during the mid-1990s, when there was an official tendency to separate out 'economic necessity' and the 'luxury' of culture. However, watching Singapore from a distance – as the 'lens' metaphor suggests – Peterson's direct engagement with the performances recedes before the more orderly and easily accessible wealth of written and electronically available information about their context. Ultimately, he succeeds in granting political legitimacy to theatre as cultural production at the expense of articulating what the theatre-makers about whom he writes actually produced.

Similarly, in *Staging Nation: English Language Theatre in Malaysia and Singapore*, the Malaysian-Australian academic Jacqueline Lo describes her chosen plays as 'paradigmatic examples of a politicized theatrical renaissance in the mid-1980s' (2004: 5), and her analytical focus as 'the ways in which identities are played out within the complex ensembles and discursive flows that produce a multiplicity of subject positions which enable us to survive and even work towards changing our society through various modes of resistance, subversion, compliance, complicity and oppositionality' (ibid.: 7). As the terminology here and throughout the book demonstrates, Lo is an assiduous student of Michel Foucault and associated writers, and her analysis consistently demonstrates the ability of the plays both to teach and enact 'the negotiation between and within contending power discourses as a necessary condition to subject formation' (ibid.: 189). In our view, the writer's distance from the encounter with performance has once again given rise to a compensatory reliance on more readily available material: in this case, poststructuralist theories of the subject.

In this light, it is unsurprising that the work of the most internationally exposed Singaporean theatre-maker – Ong Keng Sen – has elicited the most theoretically nuanced international scholarship. Much focuses on how Ong's work reproduces (Bharucha, 2000 and 2004), problematizes (Grehan, 2000 and 2001; Wee, 2004) or refreshes (Rae, 2005; Yong, 2004 and 2006) dominant paradigms of intercultural performance theory, to the extent that Ong has become a textbook reference point (Schechner, 2006: 308–10) in debates about theatre-makers as agents of global processes. Furthermore, as this proliferation of analyses in international refereed journals demonstrates, where this doubling of aesthetic form and context of inquiry extends into the theoretical concerns of the writers themselves, increased academic knowledge production on a global scale is itself an outcome.

Local knowledge as cultural production

If state performativity has proved a tempting diversion for researchers publishing for an international readership, a common characteristic of locally produced research is its embeddedness. We call this second response to the demand to 'say as I do,' the 'cultural production' model, since the resulting research is best understood as a discursive extension of ongoing creative and curatorial practices. Diverse and partisan, it serves a range of self-interested commissioning agencies, including artists who have initiated critical discourse around their own work. For example, the theatre company The Necessary Stage (TNS) launched the journal *focas: Forum on Contemporary Art and Society* (now published independently by its original editor, Lucy Davis), and two collections of essays (Necessary Stage, 1997; Tan and Ng, 2004). Artists have also published the bulk of local playscripts, many with informative introductions (Kuo, 2000 and

2003; Wong, 2005 and 2006). Meanwhile, arts spaces (Esplanade, 2004; Substation, 2001), national institutions (Oon, 2001), and government bodies (Kwok, Mahizhnan, and Sasitharan, 2002) have used the publication of research to promote their various aesthetic and programmatic agendas and to fulfill their public remits.

A positive outcome of continuity between creative practice and critical reflection is that writers are answerable both to the work and the artists, and their analyses are charged with contributing meaningfully to understanding the arts in Singapore. However, in contrast to state-mandated discourse, an infrastructure that might ensure consistency and follow-up is lacking. Local distribution is erratic, print-runs small, and writers often unaware of what has been said before. As ad hoc and often event-linked byproducts of performance, such publications are often perceived to play a commemorative, archival, or testimonial role, and their value as research diminishes accordingly. Exacerbated by an absence of critical mass, relative to the resources required to produce them, the outcomes lack efficacy.

The social science approach and performance indicators

By contrast, efficacy is arguably the watchword for that prevalent strand of performance research by Singaporean academics which derives from a social sciences model. In the work of sociologists Terence Chong (2004 and 2005) and Chua Beng Huat (2003), cultural geographers Lily Kong and Brenda Yeoh (2001 and 2003: 162–200), anthropologist Yao Souchou (2007) and, most notably, in the resolutely rational oeuvre of literary-cultural studies writer C. J. Wee Wan-ling (2002, 2003, and 2004), studied disinterest largely divests 'say as I do' of its interpellative force. Methodologically exact, stylistically restrained, and based on normatively defined frames and terms of reference, these researchers identify socially significant performances, and benchmark them against wider socio-political trends. Yet their writings are often delimited by a 'performance indicator' approach to performance that displaces cultural metaphors with socio-political metonymies.

One notable strength of this social science paradigm model of research is its breadth of inquiry. Moving between different kinds of performance – from popular forms to theatre to national and regional spectacles – and between different interpretations of performance and performativity, it more closely resembles the approach taken in performance studies than that of most local arts-focused writers.

In particular, Wee – a co-editor of the present volume – has pursued a series of thematic analyses of Singaporean theatre and pop culture that represents a rare, sustained investigation into local cultural production as continuous with the political and economic discourses of Asian modernity. Assuming the givenness of whichever performances he addresses as expressions of an emergent socio-cultural condition, he coolly renounces the consolations both of theory and of partisan belonging. As yet, however, he seems unable to reconcile his comments on the instrumentalization of aesthetic experience with a response to aesthetic effects in anything other than an instrumentalizing register that echoes his contextualizing sources. We only know Wee watched the performances he discusses because we saw him there, and although Singapore sociologist Chua Beng Huat has observed that 'outsiders don't seem to have a problem reading Singapore, it is the ones who live here who have problems reading this place' (cited in Lee, 1998: 78), we contend that Wee's dispassionate professionalism is a problem for all of us. That he, a 'local,' is able to write his embodied presence out of his analyses (or to regionalize his subjectivity) causes us to levy this critique while also questioning why we – an American (Langenbach) and a Briton (Rae) – are *unable* to do so. It signals a problem of positionality to which the issue of ethnic-national authority and authenticity is a contributing, but not defining, factor.

Although stylistically and even ideologically dissimilar, our respective publications have methodologically prefigured the present chapter through a 'thick' approach to contextualizing performance aesthetics politically and historically and in daily life (see, for instance, Langenbach, 1996, 2003, and 2006; Rae, 2004, 2006, and 2007). By dint of our involvement as 'cultural producers,' we have been selected by the editors to (re)present a ground-level 'view from Singapore' – an ironical mandate in light of Wee's 'authentic' Singaporean credentials, here reprising his 'performance indicator' role.

At the same time, like other Euro-American writers, we also privilege state performativity – albeit leavened by a focus on the socio-historical context, and contingent, embodied presence (indeed, it is this sedimentation of discourses that has led us to our central contention that what is at stake in this survey is the relationship between the subject and object of 'say as I do'). But the 'thick' approach doesn't exempt us from some of the same critiques we have made of others: if anything, it intensifies them. While all three models identified above have their strengths, we suggest that none of them – whether internationally produced theoretical perspectives, local research commissions, or 'objective' academic scholarship; whether 'thick' or 'thin' analyses – has yet proved capable of modeling an appropriate relationship between cultural performances and their researchers in the context of Singapore's performative autogenesis. Categorizing the problem does, however, suggest where we might find potential lines of development, and it is to those that we now turn.

Three new directions for research

Nuancing the nation-state: the parochial and the provincial

As we note above, great emphasis is placed on nation-building in Singapore, and a prominent strand of research has focused, deliberately or otherwise, on state performativity. By contrast, several recent publications signal useful directions for future inquiry by avoiding the exceptionalism that the analysis of Singapore's unique qualities so often entails, in favor of some determinedly unremarkable contexts for understanding performance. Margaret Chan's *Ritual is Theatre, Theatre is Ritual:* Tang-ki: *Chinese Spirit Medium Worship* (2006), is among the first locally published performance studies books, and while the grandstanding title is deceptive, the study itself provides a detailed overview of a dynamic Hokkien-Chinese-Singaporean tradition. As such, it represents a valuable contribution to the local literature on performance, and is a salutary reminder of how linguistically and culturally segregated much academic knowledge production in Singapore continues to be. Alongside Vineeta Sinha's anthropological study of ritual innovations in Singaporean Hindu Muneeswaran worship (2005: 188–235), Chan's study signals how much legwork remains to be done in taking a clear-eyed and culturally competent look at the range of performance practices that currently constitute not only the 'local' in Singapore, but the parochial – that stubbornly countervailing force in society, all too often overlooked in the aspirational striving for a globalized cosmopolitanism of the government, artists, and theorists alike.

In a related development, *Between Tongues: Translation and/of/in Asia* (2006), edited by Indonesianist Jennifer Lindsay while based in Singapore, is a rare attempt to contextualize local performance regionally. Combining academic essays and artists' perspectives on the place of translation in performances from Indonesia, Thailand, Malaysia, India, and Singapore, the collection *provincializes* Singapore, and reminds readers of the immense linguistic and cultural complexity of the South and Southeast Asian regions. While English-proficient Singapore was able to sidestep this regional complexity through its early economic appeal to Euro-American multinational corporations, it is destined to remain within this geo-social milieu and can only benefit from addressing its cultural dynamics with greater sensitivity and nuance.

Research-based practice as critical cultural production

A second promising avenue of research eschews conventional academic protocols in favor of practice-based inquiry; thereby representing one solution to the limitations of research as cultural production outlined above. Crossing languages and media, numerous critical practices have emerged in recent years that incorporate research both as process and outcome. There are various reasons for this, but given Singapore's rapid transformation into a 'Knowledge-Based Economy' – characterized, in part, by the commodification and manipulation of information as a productive activity in its own right – it is unsurprising that several artists have either reproduced, nuanced, or critiqued the aesthetics of knowledge production in their work.

In the theatre, the 'reproduction' approach has been most pronounced in Ong Keng Sen's long-term intercultural theatre project.[2] Research is integral to this project, and also to several key performances. At its best, this results in a kind of intercultural pedagogy that simultaneously enacts instruction and discovery: criteria of interpretation are established within and by the participants, even as the work unfolds. While the scope and sheer ambition of Ong's work is impressive – genocide, trauma, cultural memory, power relations, identity politics, and globalization are all recurring themes – this practice, perhaps inevitably, fails to cleave consistently to the rigors of a research agenda. It becomes problematic, however, in Ong's directorial failure to establish a framework within which research could be one of *several* judiciously mediated contexts. For while Ong's inter-Asian research aesthetic is an apposite representation of Singapore's aspirations towards being a regional 'knowledge hub,' we contend that he has yet to discover an appropriately multiplex metaphorical framework for the coherent presentation of such work. And with no let-up in his furious work-rate and prodigious productivity, the question arises whether this 'research aesthetic' is merely a pragmatic byproduct of his new projects. If so, an 'appropriately multiplex framework' is unlikely to be forthcoming.

A second strand of artist-produced research – one broadly describable as nuancing local approaches to performance – can be found in work by two of Singapore's best-known independent film-makers, Royston Tan and Tan Pin Pin. Contrasting in style and approach (Royston Tan's shorts and features are notable for their brash visual energy, Tan Pin Pin's documentaries for their observational restraint), it is nevertheless telling that both have not only featured performance in their work, but demonstrate a performative sensibility in their practice.

Royston Tan brings a theatricality to his films, which is often ampli-fied via an adroit use of performance forms that, while popular, remain under-represented in the mainstream media. His 2003 feature *15: The Movie*, a visceral portrait of underclass *ennui*, was punctuated by vulgar-ity laden Hokkien-Chinese dialect song and dance routines, delivered to-camera by teenage gangs; *15* was heavily cut by the censors, leading in part to his short film *Cut* (2004), a campy musical satire on censor-ship regulations. His 2007 feature *881*, shot during the 2006 'Hungry Ghost' festivities, told the story of two young women struggling to suc-ceed as pop singers in the *Getai* industry.[3] Most significant in the present context, however, is *Sin Sai Hong* (2006), named for the Hokkien street opera troupe – one of Singapore's oldest, and last – whose performers are featured. Commissioned by the National Museum of Singapore, Tan spent a year as an ensemble member to research the film, and the result is a lyrical and revealing 40-minute meditation on the form. Appar-ently recognizing in the flamboyant costumes and make-up a match for his own theatrical sensibility, he exercises uncharacteristic restraint by filming almost entirely in mid-shot. The camera pans slowly back and forth before the resplendent performers, who remain almost static as they sing their way through a repertoire that spans the twentieth century. Highly cinematic, it ingeniously captures the heartfelt, ostenta-tious, worldly-wise qualities of the form that would arguably evade any more conventional attempt at theatre documentation. In so doing, it enhances another significant characteristic of the opera: as a repository for embodied cultural memory.

Tan Pin Pin also focused on local Chinese folk culture in her doc-umentaries *Moving House* (2001) – about mass exhumations mandated by the state to make way for housing developments – and *Gravedigger's Luck* (2003) – about the precautions a gravedigger takes against the bad luck that such exhumations are said to bring. This extended to a more wide-ranging concern with everyday performance in *Singapore GaGa* (2005), an impressionistic portrait of human sound in the city-state. Suturing together, among others, elderly communists singing revolu-tionary songs, Arabic-speaking cheerleaders at a Malay-Muslim *madrasah* sports day, a hymn-warbling wheelchair-bound tissue seller, and toy piano virtuoso Margaret Leng Tan performing John Cage's *4′ 33″* in a government-built public-housing estate, Tan draws attention to the role of performance in mediating both social and spatial relationships. Moreover, her unassuming documentary persona appears to draw these performances out of her subjects in ways that are both matter-of-fact and compelling. However, in foregrounding performance, we aver that Tan obscures context – of class, of social attitudes towards the disabled, of mental illness, and of how the people depicted in the film become exoticized and isolated from their milieus. It appears that one of the hardest things to achieve in a state which demands 'say as I do,' is to tell it like it is.

A third, more explicitly critical strand of practice-based research in Singapore derives from the visual arts. Drawing on a largely suppressed legacy of politically engaged aesthetic exploration, it pitches questions of history and identity against an economic process prone to homogenization and amnesia. Some artists, such as Amanda Heng and Ho Tzu Nyen, use performance to recreate and re-embody aspects of this history, albeit in altered and contestatory form. Here, however, we focus on the curatorial practice of Koh Nguang How, which uses performativity as a means of suturing the gaps between past and present.

In 2004, Koh presented an installation/performance entitled *Errata: Page 71, Plate 47. Image Caption. Change Year: 1950 to Year: 1959; Reported September 2004 by Koh Nguang How*.[4] Starting from an error in the dating of a social-realist painting, Chua Mia Tee's *National Language Class*, in Singapore Art Museum director Kwok Kian Chow's authoritative survey book, *Channels & Confluences: A History of Singapore Art* (1996), Koh turned the independent Singapore gallery p-10 into an archival depository, detailing the socio-political and artistic contexts of the work, with Koh on hand to assist visitors in understanding the personal, social, and ideological links between them. However, rather than build a comprehensive picture of the period, the peculiar effect of this ground-up, ad hoc history lesson was an increasingly powerful sense of how much one did *not* know – not only for want of facts, but because there are certain things that cannot yet be known about that period. The demonstrable 'non-finality' of Koh's research is in stark contrast to the national tendency to gravitate toward a final, authoritative historical state discourse. His Singapore is an open, perennially unfinished project where most of the historical iceberg remains hidden beneath the surfaces of the everyday. As such, his ongoing project of genealogical retrieval elucidates a particular performative research methodology that continues to examine the Singapore 'object of research' as a heterogeneous anti-foundational disturbance at the margins of official historicism.

Beyond performance indicators: integrating research practices

While it is clear that research-based practice is not meant to be a comprehensive substitute for a critically informed and methodologically rigorous research culture, we take the extent of numerous Singapore artists' critical engagement and inquiry-based practice to be both a challenge to, and an opportunity for, rethinking the 'performance indicator' model of local research and the presumption of strong-state analysis that it implies. Practitioner-researchers present a particular challenge to writers operating within more conventional research structures; but they also represent a long-standing feature of much East Asian performance, where aesthetic 'theories' are traditionally embedded in and interrogated by means of practice, rather than existing separately, in written form. A local 'ecology of performance research,' even in post-industrial Singapore, is not reducible to the output of the academy and related agencies.

In this light, it is significant that the tenth annual conference of Performance Studies international (PSi), held in Singapore in June 2004, was organized independently by a consortium of locally based theatre-makers, writers, and administrators. Seeking to mark a decisive moment in the globalization of Performance Studies as a discipline, as well as reconsidering the place of Singapore in the dynamics of regional performance knowledge production, the organizers (including the authors) first held a regional pre-conference in Penang, Malaysia in 2003 to survey the field and establish the principles that would guide the formulation of the actual Singapore event. In 2006, another regional roundtable, 'Panic Buttons: Crisis, Performance, Rights,' was held in Kuala Lumpur, Malaysia, where artists, activists, and academics from Singapore, Malaysia, Indonesia, and Thailand affirmed their commitment to developing a network for dialogue and exchange in the midst of a climate marked by increasing threats from a range of political and civic agencies to regional cultural production. Although limited to a largely English-literate group in the case of 'Panic Buttons,' we would argue that in bringing together a range of social, intellectual, and creative actors, such events are probably more representative of the regional situation vis-à-vis performance knowledge and research than more specialized or professional institutions such as universities are currently capable of. It offers one way of addressing the extent to which local knowledge is still communicated orally, or resides within embodied experience, even in apparently advanced post-industrial economies.

Conclusion

Singapore is distinctive and unique, but it is not, despite the state's ingenious ideological operations, singular; nor can it be detached from its historical, geographical, or political contexts. While it is tempting to dismiss such a statement as little more than a truism, the fact remains that we have yet to see the development of attitudes and research approaches that can account sufficiently for the nuanced interpretations the situation demands.

As we hope to have indicated, cultural performance research in Singapore is wide-ranging, layered, and paradoxical in its objects of inquiry, methods of production, languages of expression, frameworks for analysis, sources of funding, and opportunities for reception. We have therefore identified three characteristics of local research and its contexts – namely the extent to which researchers represent and respond to state performativity, cultural production, and the demand to 'indicate' or measure performance – and have noted numerous avenues for further development: rethinking the nation in both local and regional terms; research-based creative practice; and hybrid events that gather a representative range of stakeholders in the 'ecology' of local and regional research.

However, the inevitability of such developments cannot be assumed. Cultural performance research remains tied to the intentions and fate of researchers and their objects of study, and is subject to political constraints and the vagaries of institutional imperatives. And because these vagaries still inflect our views of Singapore performance, maintaining a discreet distinction between state desire and the desires of the artist-researcher is impossible. The reflexive analysis of institutional performance from within remains substantially constitutive of Singapore performance studies. Developing performance research, then, will be as much a matter of acknowledging the demands of these variables as of producing more accurate and perceptive analysis of the diverse range of events and processes that constitute the field.

Notes

1. One of the most densely populated nations in the world, at independence Singapore was 581 sq. km in size, and had a population of 1.9 million. By 2006, the population had grown to almost 4.5 million, and, as a result of an aggressive land reclamation policy, the island was 697 sq. km. According to the racialized categorizations used by the government, the population was approximately 75 percent Chinese, 14 percent Malay, 9 percent Indian, and two percent 'Other.'
2. For details, visit the TheatreWorks (Singapore) official website at: http://www.theatreworks.org.sg.

3. These popular concerts are held all over Singapore on temporary stages during the Taoist Hungry Ghost month (around August), to appease spirits and entertain mortals.
4. The installation was hosted by the independent art space p-10, under the curatorship of Jennifer Teo and Woon Tien Wei. The latter's work as member of the new media collective, <tsunamii.net>, also figures as a significant sign in this constellation of artists working on the border of theory-qua-praxis. See <tsunamii.net> for more details.

References

Bharucha, R. *Consumed in Singapore: The Intercultural Spectacle of Lear* (Singapore: Centre for Advanced Studies, National University of Singapore, 2000).
———. 'Foreign Asia/Foreign Shakespeare: Dissenting Notes on New Asian Interculturality, Postcoloniality, and Recolonization,' *Theatre Journal*, 56:1 (2004) 1–28.
Buchanan, I. *Singapore in Southeast Asia: An Economic and Political Appraisal* (London: G. Bell & Sons, 1972).
Chan, H. C. 'Politics in an Administrative State: Where has the Politics Gone?' (1975), in *Understanding Singapore Society*, ed. J. H. Ong, C. K. Tong, and E. S. Tan (Singapore: Times Academic Press, 1997), pp. 294–306.
Chan, M. *Ritual is Theatre, Theatre is Ritual:* Tang-ki: *Chinese Spirit Medium Worship* (Singapore: SNP Reference, 2006).
Chong, T. 'Mediating the Liberalisation of Singapore Theatre: Towards a Bourdieusian Analysis,' in *Ask Not: The Necessary Stage in Singapore Theatre*, ed. C. K. Tan and T. Ng (Singapore: Times Editions-Marshall Cavendish, 2004), pp. 223–48.
———. 'From Global to Local: Singapore's Cultural Policy and Its Consequences,' *Critical Asian Studies*, 37:4 (2005) 553–68.
Chua, B. H. *Communitarian Ideology and Democracy in Singapore* (London and New York: Routledge, 1995).
———. *Life is Not Complete without Shopping: Consumption Culture in Singapore* (Singapore: Singapore University Press, 2003).
Esplanade, The. *Coping With The Contemporary – Selves, Identity And Community* (Singapore: The Esplanade, 2004).
Gibson, W. 'Disneyland with the Death Penalty,' *Wired*, 51–5 (September/October 1993) 114–16.
Grehan, H. 'Performed Promiscuities: Interpreting Interculturalism in the Japan Foundation Asia Centre's *LEAR*,' *Intersections*, 3 (2000).
———. 'TheatreWorks' *Desdemona*: Fusing Technology and Tradition,' *The Drama Review (TDR)*, 45:3 (2001) 113–25.
Kong, L. and B. Yeoh. 'The Construction of National Identity through the Production of Ritual and Spectacle: An Analysis of National Day Parades in Singapore,' in *Singapore*, ed. G. Rodan (Aldershot, England: Ashgate, 2001), pp. 375–402.
———. *The Politics of Landscapes in Singapore: Constructions of a 'Nation'* (Syracuse, NY: Syracuse University Press, 2003).
Koolhaas, R. 'Singapore Songlines: Portrait of a Potemkin Metropolis ... or Thirty Years of Tabula Rasa,' in *S, M, L, XL*, ed. R. Koolhaas and B. Mau (New York: Monacelli Press, 1995), pp. 1008–89.

Krishnan, S. 'What Art Makes Possible: Remembering Forum Theatre,' in *Nine Lives: 10 Years of Singapore Theatre – Essays Commissioned by The Necessary Stage*, ed. The Necessary Stage (Singapore: The Necessary Stage, 1997), pp. 200–11.

Krishnan, S., S. Kuttan, W. C. Lee, L. Perera, and J. Yap. (eds) *Looking at Culture* (Singapore: Artres Design and Communications, 1996).

Kuo, P. K. *Images at the Margins: A Collection of Kuo Pao Kun's Plays* (Singapore: Times Publishing, 2000).

———. *Two Plays by Kuo Pao Kun: Descendants of the Eunuch Admiral and The Spirits Play*, ed. C. J. W.-L. Wee and C. K. Lee (Singapore: SNP Editions, 2003).

Kwok, K. C. *Channels and Confluences* (Singapore: Singapore Art Museum, 1996).

Kwok, K. W., A. Mahizhnan, and T. Sasitharan. (eds) *Selves: The States of the Arts in Singapore* (Singapore: National Arts Council, 2002).

Langenbach, R. '*Leigong Da Doufa*: Looking Back at Brother Cane,' in *Looking at Culture*, ed. S. Krishnan, S. Kuttan, W. C. Lee, L. Perera, and J. Yap (Singapore: Artres Design and Communications, 1996), pp. 123–38.

———. *Performing the Singapore State: 1988–1995*, Unpublished PhD Dissertation, Centre for Cultural Research, University of Western Sydney, Nepean, Australia (2003).

———. 'Jacked Off with No Pleasure: Censorship and the Necessary Stage,' in *Ask Not: The Necessary Stage in Singapore Theatre*, ed. C. K. Tan and T. Ng (Singapore: Times Editions-Marshall Cavendish, 2004), pp. 201–22.

———. 'Garlands of Love: Socialist Realism in Singapore,' in *Eye of the Beholder*, ed. J. Clark, M. Pelleggi, and T. K. Sabapathy (Sydney: Wild Peony Press, 2006).

Lee, K. Y. 'Transcript of a Press Conference Given by the Prime Minister of Singapore, Mr Lee Kuan Yew, at Broadcasting House, Singapore, at 1200 Hours on Monday 9th August, 1965,' http://stars.nhb.gov.sg/stars/public/index.html (accessed 21 August 2009).

Lee, W. C. 'Time, Landscape and Desire in Singapore,' in *Singapore: Views on the Urban Landscape*, ed. Lucas Jodogne (Antwerp: Pandora, 1998), pp. 54–80.

Lindsay, J. (ed.) *Between Tongues: Translation and/of/in Performance in Asia*. (Singapore: Singapore University Press, 2006).

Lo, J. *Staging Nation: English Language Theatre in Malaysia and Singapore* (Hong Kong: Hong Kong University Press, 2004).

Necessary Stage, The. (eds) *Nine Lives: 10 Years of Singapore Theatre – Essays Commissioned by The Necessary Stage* (Singapore: The Necessary Stage, 1997).

Oon, C. *Theatre Life! A History of English Language Theatre in Singapore through* The Straits Times *(1958–2000)* (Singapore: Singapore Press Holdings, 2001).

Peterson, W. *Theater and the Politics of Culture in Contemporary Singapore* (Middletown, CT: Wesleyan University Press, 2001).

Phillips, J. 'Singapore Soil: A Completely Different Organisation of Space,' in *Urban Space and Representation*, ed. M. Bradshaw and L. Kennedy (London: Pluto Press, 2000), pp. 175–95.

Rae, P. ' "10/12": When Singapore Became the Bali of the Twenty-First Century?' *focas: Forum on Contemporary Art and Society*, 5 (2004) 222–59.

———. 'Don't Take It Personally: Arguments for a Weak Interculturalism,' *Performance Research*, 9:3, 'On Civility' (2005) 18–24.

———. 'Why There is Wind: Power, Trees, Performance,' in *A Performance Cosmology: Testimony from the Future, Evidence of the Past*, ed. J. Christie, R. Gough, and D. Watt (London and New York: Routledge, 2006), pp. 207–14.

————. 'Cat's Entertainment: Feline Performance in the Lion City,' *The Drama Review (TDR)*, 51:1 (2007) 119–37.

Schechner, R. *Performance Studies: An Introduction*, 2nd edn (London and New York: Routledge, 2006).

Seet, K. K. 'Interpellation, Ideology and Identity: The Case of *Talaq*,' *Theatre Research International*, 27:2 (2002) 153–63.

Sinha, V. *A New God in the Diaspora? Muneeswaran Worship in Contemporary Singapore* (Singapore: Singapore University Press, 2005).

'*Talaq* Documents,' *focas: Forum on Contemporary Art and Society*, 1 (2001) 181–210.

Tan, C. K. and T. Ng. (eds) *Ask Not: The Necessary Stage in Singapore Theatre* (Singapore: Times Editions-Marshall Cavendish, 2004).

Substation, The. *Open Ends: A Documentation Exhibition of Performance Art in Singapore* (Singapore: The Substation, 2002).

Vacca, G. 'Intellectuals and the Marxist theory of the State,' in *Approaches to Gramsci*, ed. A. S. Sasoon (London: Writers and Readers Cooperative Society, 1982), pp. 37–69.

Wee, C. J. W.-L. 'Bland Modernity, Kitsch and Reflections on Aesthetic Production in Singapore,' *focas: Forum on Contemporary Art and Society*, 3 (2002) 118–33.

————. 'Creating High Culture in the Globalized "Cultural Desert" of Singapore,' *The Drama Review (TDR)*, 47:4 (2003) 84–97.

————. 'Staging the Asian Modern: Cultural Fragments, the Singaporean Eunuch, and the Asian Lear,' *Critical Inquiry*, 30 (2004) 771–99.

Wee, C. J. W.-L., and C. K Lee. 'Breaking Through Walls and Visioning Beyond – Kuo Pao Kun Beyond the Margins,' in *Two Plays by Kuo Pao Kun: Descendants of the Eunuch Admiral and The Spirits Play*, ed. C. J. W.-L. Wee and C. K. Lee (Singapore: SNP Editions, 2003), pp. 13–34.

Wong, E. *Invitation to Treat: The Eleanor Wong Trilogy* (Singapore: Firstfruits, 2005).

————. *Earlier* (Singapore: Firstfruits, 2006).

Yao, S. *Singapore: The State and the Culture of Excess* (London and New York: Routledge, 2007).

Yong, L. L. 'Ong Keng Sen's *Desdemona*, Ugliness, and the Intercultural Performative,' *Theatre Journal*, 56 (2004) 251–73.

————. 'Shakespeare and the Fiction of the Intercultural,' in *A Companion to Shakespeare and Performance*, ed. B. Hodgdon and W. B. Worthen (Oxford: Blackwell, 2006), pp. 527–49.

9
The Performance of Performance Research: A Report from Germany

Sibylle Peters

Performance as and of research, research as and of performance

In his *Introduction to Performance Studies* – a book that does not refer to any current German scholar – Richard Schechner (2002) begins by claiming that performance studies resists definition as a field, and yet he goes on to provide precisely that. Similarly, in the case of the present volume, which aims to put into focus how performance research has developed *beyond* its major representation in the Anglo-American academic context, any attempt to draw a map of performance research in a certain country performs itself an act of representation and engages in the construction of an identity or, at least, an entity – in this case that of German performance research. This raises at least two issues. Though performance studies as it is emerging in Germany may not be fully acknowledged as a part of the Anglo-American academic discourse, it certainly does not view itself in opposition to it either. And secondly, the already problematic construction of any kind of national identity is an even more questionable project in the case of Germany, when we take into account its history of nationalism.

To avoid these pitfalls I would like to turn the task I am confronted with into an opportunity for doing something I have wanted to pursue for a while, namely to develop a provisional – and admittedly somewhat sketchy – typology of 'performance research,' based on local phenomena taken from the performance scene and the academic discourse in Germany as I witnessed them in the decade from the mid-1990s to around 2006. This typology does not claim to be an account of what German performance research *is* but attempts to reflect on the net of relations that exists between performance and research from the perspective of a scholar and performer based in Hamburg and Berlin and working between academia and performance practice.

The greatest challenge performance research poses derives from the fact that by definition it transcends the boundaries of academic work, and that performance itself may be viewed and practiced as research. Thus the terms 'performance' and 'research' are flexible in their relation to each other. Performance is not only the object of research, but it can also provide a range of methods for research or formats in which the outcomes of research may be presented. This raises the question under what conditions performances might be called research. But it also asks what kind of performances are in turn constitutive for research. In short – what makes research a possible paradigm of performance, and what makes performance a configuration of research? The following typology will be divided into three main parts: research on performance (artistic and cultural), performances of research (transitions between performance and research), and performance-as-research (on performance).

And whilst I will highlight what appears to me to be a local particularity, certain shapes of performance research might well be detected by attending to the typology's idiosyncratic features, that is, by considering what seems to be neglected or taken for granted – in other words, by reading it as symptomatic of its local origins.

Research on performance

'Performance' is, obviously, not a German word. The German term 'Performanz' is only a partial translation, referring as it does exclusively to the theory of language (where it denotes the opposite of 'competence'). Nevertheless, 'performance' is now included in the *Duden*, the leading German dictionary, which differentiates between its general use in English to refer to a 'staging' (usually translated as 'Aufführung' or 'Vorführung') and its more specific use in German, which the dictionary identifies as referring to 'einem Happening ähnliche künstlerische Aktion' ('an artistic action akin to a happening') (*Duden*, 1996). Thus the word 'performance' in German is used primarily to indicate an artistic action (i.e., a work of performance art) that differs from traditional theatrical performances. To call the performance of a play, of a lecture, of music, or dance with the English term a 'performance' in German already indicates that the performance was designed in some special way – thus we may say that in German, the term 'performance' already implies a reference to the performance dimension of a performance. Moreover, a dancer, musician, or actor would not usually be called a 'performer,' unless to hint at a different and less formal artistic qualification.

This difference in meaning might suggest a potentially different starting point for research on performance in the German context. But interestingly, leading scholars have tended to neglect this aspect. Erika Fischer-Lichte's influential *Ästhetik des Performativen* (2004) could even be read as an attempt to compensate for this shift in meaning. With the wider sense of the English word 'performance' as a theatrical 'staging' in mind, Fischer-Lichte argues that modern art in general is characterized by a 'performative turn.' Performance aspects have become increasingly important in all forms of art and have thereby encouraged interdisciplinary approaches that have questioned the boundaries between them. This performative dimension is defined by Fischer-Lichte – and this seems to be the general consensus across the field – by the twists and turns in the relationship between artist and audience, and in the relationship between materiality and meaning, which has challenged the hegemonic position of the latter. Furthermore, the performative dimension presupposes that space and time have become artistic media in their own right.

From this perspective, examples of performance art are regarded as resulting from a general sense of staging that has spread through all forms of art (see Fischer-Lichte and Roselt, 2001). A definition like this keeps the stage performance as a template in mind. Thus it implicitly claims that theatre studies has a certain competence to deal with developments in other forms of art, even if the artists which are mostly cited in this context (Joseph Beuys, Marina Abramović, Viennese Actionism, Fluxus) clearly belong to the tradition of fine art.[1]

That strategic considerations are of some importance here may be indicated by the fact that Fischer-Lichte is the director of an interdisciplinary research program, entitled *Kulturen des Performativen* (Cultures of the Performative). It is important to mention that the interdisciplinary discourse that defines performance research in Germany preceded the efforts to institutionalize something like disciplinary performance studies. From 1994 onward, extensive public grants had been given to a prior research program (involving projects from all disciplines within the humanities and again chaired by Fischer-Lichte) that was concerned with 'theatricality' and dealt with 'performance' and 'performativity' as key terms. Its successor, *Kulturen des Performativen*,[2] which is based at the Freie Universität in Berlin, is still working successfully, again involving a number of research projects from different disciplines.

One reason for this success may be that the program does not claim to execute the often-evoked 'performative turn' whilst switching from text to performance as a new paradigm, but it makes 'performative turns' in their particular relation to text its very object of research – the network deals primarily with the relation between performativity and textuality within European history. Thus, the program dissociates itself from the momentum of an actual performative turn within the humanities by relocating it into a number of historical scenarios.

Another condition for the success of the program is that 'performance' had already become a crucial term in many disciplines, such as sociology, ethnology, historical, and cultural studies. Here the primary references are taken mainly from the Anglo-American discourse: Singer, Goffman, and Schechner are important sources in this context, as is linguistic theory and speech act theory, and recently, in particular, the work of Judith Butler. These scholars are generally widely translated and read, and the corresponding discussions in a German context do not divert too much from their Anglo-American origin when it comes to topics like 'ritual' or 'gender.' Thus I am going to limit myself to naming a few major focal points of performance research which seem to differ in some respects from their Anglo-American predecessors.[3]

As performance is no longer primarily defined by the presence of a human performer, but rather by a shift in the relation between materiality and meaning and between actors and audiences, performance studies in Germany has moved closer to media studies. One point of contact has been the recent, vividly conducted discussion about the voice. Berlin-based philosopher Sybille Krämer has outlined how 'performativity' can be read as 'mediality' and vice versa in this context (Krämer, 1998). Informed by phenomenological concepts of body and perception and Benjamin's notion of 'aura,' philosopher Dieter Mersch, based in Potsdam, has theorized the relationship between mediality and performativity by drawing on examples of (post)modern fine art (Mersch, 2002). Reading Mersch and Krämer it becomes clear that theories of performance are attractive not least because (by referring to Austin, Wittgenstein, Nietzsche, or poststructuralism) they maintain contact with philosophies of language whilst simultaneously transcending language as the main philosophical theme of the last century.[4]

In dealing with this transition one concept has turned out to be of particular interest – the concept of 'Figur' and 'Figuration,' which in German applies to figures of speech as well as to characters on stage. 'Figur/ation' carries an ambivalence that blurs dual concepts of mind and body – it is situated between media and, as a concept, tries to grasp what escapes systematic structures (see Brandstetter and Peters, 2002).

Finally, although this is of rather marginal concern to my present discussion, I would like to add that – read with German history in mind – theories of performance raise certain questions concerning the problematic mystification of the act (or the deed – German: *die Tat*), its world-creating power and bodily presence that can be found in German philosophy since the era of Idealism. To mark the difference between these traditional figurations of thought and contemporary concepts of performance is an important task, which has led the philosopher Werner Hamacher, for example, to replace the term 'performative' with the term 'afformative' (Hamacher, 1994).

Performances of research

Whilst the textual paradigm provided a continuum between the humanities, their primary medium and the objects of their research, this is no longer the case for performative phenomena. Thus, the shift from the textual to the performative has the potential to renew the German scholarly tradition. Considering the rather strict separation between theory and practice within this tradition, it could be expected with some justification that this shift would open a passage from scholarly to artistic practice by diversifying theoretical practices, which are no longer confined to the production of textual content. But, although it cannot be denied that in the context of academic research a lot of methodological reflection on performance practice has been accomplished in recent years, the search for new approaches has not been particularly successful when it comes to academic practice itself – yet. When the humanities in general are put under growing pressure to legitimize themselves economically, this may not be the best time for experiments. Still, performance research as a field cannot evolve without exploring the possible transitions between the research on performance, the performances of research and performances as research.

There have been some efforts to reflect on the performance of research worth mentioning. One was initiated by Gabriele Brandstetter, Professor of theatre and dance studies in Berlin, and her collaborator Hans-Friedrich Bormann. Contrary to the definition of performance as ephemeral and thus impossible to record and reproduce that dominated much of performance studies' thinking in the 1990s, Brandstetter and Bormann argued that, like so many other binaries, performance also deconstructs the opposition between event and documentation (Brandstetter, 2004; Bormann and Brandstetter, 1999; Bormann, 2003). From early body art to choreographic experiment, from video art to works which mainly exist as anecdote or rumor, performance makes us consider the (im)possibilities of documentation by making the document, the trace, the map part of the performance itself. Countering the view that scholarly reflection is always already in some radical way too late, their argument has an important consequence – as it deconstructs the difference between event and document, a performance does not pass but continues wherever it leaves traces. This makes the academic reflection that is triggered by a performance by implication a part of the performance itself. It could put an end to the discussion whether or not artistic perspectives should be a part of academic practice, as academic practice is revealed to be already implicit in the perspective of performance art.

There is also the performance of research in its own right. As Robert Felfe, a collaborator on *Kulturen des Performativen*, who has himself been concerned with the spectacular character of the experiment in the early days of natural sciences, has proposed, 'knowledge is constituted in forms of representation, staged scenes of reception, media practices and fields of social action' (Felfe, 2002, translation Peters). Governed by the term 'performativity of knowledge,' a growing interest in the culture of science is leaving the limitations of a history of pure thought behind and is paying attention to the manner, material, media, instruments, institutions, and so on that have influenced the production of knowledge (Schramm et al., 2006).[5] Nevertheless, there is a lot of work still to be done to elaborate the points of contact between science studies and performance research. Three interrelated areas are of particular importance in this regard, which all call into question the politics of knowledge production: the performance of assemblies (i.e., their constitution and their production of truth or consensus, see Latour and Weibel, 2005), the (public) presentation of knowledge and its production, and the performance of evidence (Peters and Schäfer, 2006). But to be concerned with the performativity of knowledge on the one hand and to experiment with the performance of science itself on the other are two different matters. So far, few transitions lead from one to the other that could be called 'performance research' in a wider sense.[6]

Curating and teaching performance-as-research

If the future of performance research depends largely on the development of hybrid formats that allow transitions between research on performance and performance-as-research, one important structural condition for such a development is the existence of an independent curatorial practice that strives to bring together art and theory in new ways.

Although this task is occasionally undertaken from within academic institutions, the results are often disappointing. Performances may become part of academic conferences, but they are often confined to the complementary program rather than taken as serious contributions to the academic discourse.[7] In order to change academic structures, it appears necessary to establish a new kind of audience and a different manner of attention first. This is what a number of independent curators are trying to achieve, which makes them key players in the development of a new form of performance research. Unfriendly Takeover, for example, a curators' collective from Frankfurt,[8] has invented new presentational formats which combine performance and theory.

The Summer Academy at the Mousonturm in Frankfurt has for many years brought together theory and performance in an annual program of workshops, shows, discussions, and lectures. The *Mobile Academy* by Berlin-based curator Hannah Hurtzig has held summer schools on themes such as 'The Future of Work' (Bochum, 1999) or 'The Refugee: Services rendered to Undesirables' (Berlin, 2002) in different locations.[9] Hurtzig has also developed 'academies' in miniature format: the *Schwarzmarkt für nützliches Wissen und Nicht-Wissen* (Black Market for Useful Knowledge and Non-Knowledge) functions as a marketplace for 100 experts, who pass on their knowledge to members of the audience in one-to-one encounters in exchange for a small fee.

Focusing on performance as a means of political information and manifestation in the time of globalization, the curator and theatre director Matthias von Hartz has organized a series of events which have brought together academic, artistic, and political lectures and performances that have crossed discursive boundaries. The curatorial collective of feminist artists, *thealit Frauen.Kultur.Labor*, based in Bremen, have for the past 15 years intervened in contemporary discussions on points of contact between art, science, and technology. The *Performer Stammtisch* is a network of a different kind, an association of freelance performers from Berlin,[10] who meet regularly to present work in progress or to edit archival material, to engage in 'performance battles' or re-enactments, and to reflect on the conditions of performance making. This network suggests that aspects of our urban and precarious life may be turned into objects of a collective research undertaking. As such performance research qualifies as a contemporary mode of survival.

Another prerequisite for the development of performance research is without doubt the provision of education and training. Many of the setbacks that this development has experienced in recent years have been due to the lack of the kind of double qualification that is so needed in the field. For years there was only one University program that provided training in both artistic practice and the theory of performance: The Institut für Angewandte Theaterwissenschaft (Institute for Applied Theatre Studies) at the University of Giessen, currently under the directorship of composer, Heiner Göbbels.[11]

How influential this institute has been for the development of performance research in Germany is indicated by the fact that the majority of people and networks mentioned in this chapter has in one way or other been linked to its program. Following the Anglo-American model of theatre studies, the institute's website states that 'the main interest is in theatrical research [...] on a scientific as [well as] on a theatrical-practical basis' (Institute for Applied Theatre Studies, Giessen). This, it has to be stressed again, is very much in contrast to the German tradition, which regards all disciplines of cultural studies as 'sciences' (*Wissenschaften*) and defines science and art in opposition to each other. To postulate a permeability between scientific and artistic research, therefore, means to leave behind implicit assumptions about the nature of academic (or scientific) and artistic work and to rethink fundamental ideas about both – a work that is far from done.

Whilst Germany has a strong tradition of performance art as a genre of the visual arts, no sustained cooperation as yet exists between art schools (in Germany called 'academies' of fine art) and the university-based departments of theatre studies. Again, this lack is due to the structural gap that exists between the scholarly tradition of the humanities within the framework of the university and the practical training that is provided by art schools. To bridge this gap is a challenge for the future. Nevertheless, there are and have been some attractive possibilities for studying performance art at art schools, with teachers whose work has been central to its history: Marina Abramović taught at the Hochschule für Bildende Künste Braunschweig (Braunschweig University of Art) from 1997 to 2004, and Valie Export until recently taught media art in Cologne.

With regard to education it may generally be said that opportunity and difficulty these days lie side by side. The increasing pressure to legitimize the pursuit of cultural studies within today's social and economic climate promotes the introduction of educational programs that bring theory and practice together in new ways. Yet, the same pressure also discourages scholars and artists from making research the primary link between theory and practice, instead replacing it with marketable concepts such as 'applicability' and 'competence.' To ground performance studies in a transition between scientific and artistic practices of research remains a challenge.

Performance-as-research

In a perhaps less specific but more widespread version, the notion of 'performance-as-research' is used by contemporary artists to mark their distance from a traditional view of art as the intuitive work of genius. To call it 'research' emphasizes that art is connected to more or less transparent procedures – procedures which are not primarily concerned with the inner or spiritual life of the artist, but with social, political, historical, and material issues. It also points towards an interdisciplinary exchange between artistic methods that contrasts with the notion of mastery of a specific artistic craft. Indeed, as a consequence of 'research' becoming an artistic paradigm, the traditional dualism of ingenious creation on the one hand (which is seen as 'miraculous') and mastery on the other (passed on through teaching) is being dismantled.

In the context of contemporary theatrical performance, 'performance-as-research' simply marks the most important alternative to fiction-based representational theatre. Instead of staging a drama, ensembles who call themselves 'performance groups' think of their work as a performative investigation that applies a wide range of methods. Their aesthetic is based on a simple principle. A performance-as-research project is successful if the investigation and the presentation of its outcomes, which is the performance itself, are not separate processes but permeate each other.

One example for this way of working is the performance group Showcase Beat le Mot, based in Hamburg and Berlin. Objects of their recent performative investigations have included the revolutionary history and economic present of China (*Alarm Hamburg Shanghai!*, 2005), and pirates as a model of resistance (*Europiraadid*, 2006). Whilst role playing and Brechtian narratives are still part of the work, they are just two performance tools among many, including forms of lecturing, battling, dancing, or sharing meals and drinks with the audience. 'Performance research' here denotes making theatre a part of contemporary popular culture, which might discover its affinity to comic books or DJ-culture whilst articulating a form of critique that is freely shifting between media and genres.

Research projects carried out in the format of performances might, of course, also explore performance itself, or more precisely, a certain type of cultural or artistic performance. A relatively well-known example for this kind of performance research is *perform performing – A trilogy on the sense and senselessness of looking at dance as work* by Jochen Roller (2003), a dance work which incorporates elements of lecturing. *No Money, No Love,* the first and most widely shown part of the three-part performance, deals with the economic conditions of working as a dancer, which are analyzed in the manner of a typical business presentation, complete with flip-chart. This reveals not only the absurdities inherent in viewing art from an economic perspective but also hints at the fact that the freelance performer has lately become a model for a new flexible, self-responsible worker, who remains motivated even under the most pressured conditions. As a result, *perform performing* deals with the transitions between performance as art and performance as a certain figure of productivity, which makes it more than just a self-reflexive performance about dance.

In dance, the boundaries between artistic performance and the performances of everyday life have been challenged for some time. At least two reasons spring to mind why dance has a specific affinity with performance research. Firstly, compared to the bourgeois form of theatre, the embodied performance of dance has been less dominated by fiction, narrative, and drama. And secondly, in dance the body in motion is self-evidently always already both the object and the instrument of research. Consequently, even traditional or early modern dance raise the kinds of questions otherwise associated with post-avant-garde performance art: the problems of actuality, enactment, and notation, how knowledge in dance is transmitted, and what kinds of knowledges participate in this process. It might be argued that dance is somehow predestined to undertake performance-as-research. Nonetheless, the shift of dance toward 'research' as a new paradigm cannot be understood without acknowledging the performative theory of the body, which in Germany has been discussed and elaborated as much as it has been in the Anglo-American or French context. Its importance for dance is obvious – it relates to the critique of rational discourse that is associated with contemporary dance, whilst opening up a differentiated dialogue between cultural theory and dance practice. Consequently, the coming-together of practice and theory in the field of dance is thriving – as was demonstrated by the large international dance congress that took place in Berlin in the spring of 2006, which brought together over 1600 participants under the title *Knowledge in Motion* and with the explicit aim to promote dance as a 'culture of knowledge.'[12]

Part of this enterprise are two new Masters programs which concentrate primarily on dance: one housed at the Freie Universität Berlin and founded by Gabriele Brandstetter, and one at the University of Hamburg and led by dance sociologist Gabriele Klein in cooperation with Wolfgang Sting. In Hamburg, the first intake of students began their studies in

2005, whilst in Berlin a dance lab at the Institut für Theaterwissenschaft (Department of Theatre Studies) operates as an experimental space to explore new combinations of theory and practice in the teaching of dance.

A dance performance may help to describe a further manner in which to conduct performance-as-research. The production *Warum tanzt Ihr nicht?* (Why are you not dancing?, 2004) by performance group She She Pop presented itself as a ball night, in which the audience was invited to take part with the help of a series of games. The performance thus replaced representation with participation, or, as Mieke Matzke from She She Pop put it, replaced 'the play with the game' (Matzke, 2005). This kind of performance-as-research project creates a balance between reality and fiction that allows its participants to inspect in a playful manner the rules and conventions which govern cultural performances.

That performance is able to deconstruct or at least defer the division between reality and fiction can be utilized for research in more ways than one. In certain cases this presupposes an approach to fiction that turns traditional aesthetics inside out. While bourgeois theatre is fictitious but tries to appear as 'real,' contemporary performance projects often appear fictitious but when scrutinized pass the reality-test. The performance group geheimagentur, for example, opened an *Office for the Collection of Miracles* (2005) in the city of Bochum. It attracted people who wanted to find out about the hoax, but were instead confronted with performers who pursued their task in all seriousness. In actions like these performance-as-research is combined with performance as a means of political intervention, a notion that is strongly influenced by French situationism and the notion of the 'communication guerrilla' (Blisset and Brünzel, 2001). Research and political intervention here join forces in a kind of definalized tactics, which makes reference also to de Certeau, and Deleuze and Guattari (Deleuze and Guattari, 1988). This could again be described as an inversion of traditional aesthetics. While aesthetics since Kant is defined as being free from the instrumental relations of means and goals, these performance tactics overdo instrumentality and take the goal itself for a means. What was formerly seen as the autonomy of art thus contributes to producing a 'temporarily autonomous zone' within society.[13] As a consequence of these kinds of projects, the concept of 'participation' has been linked to the development of performance-as-research in a particular way. This is exemplified by the work of Rimini Protokoll, a group who have reinvented the form of documentary theatre by involving real-life experts as performers. Their stories are arranged around a certain topic – in the project *Deadline* (2002), for example, a florist, a stonemason, a funeral speaker, a medical student, and other experts talked about their professional dealings with death.

It goes without saying that in such work the theatrical competence of the participants is incidental. Nevertheless, performances like these have led to something of a professionalization of performance work with amateurs, which is no longer primarily viewed as pedagogic. As much as this development must be welcomed, it is high time we scrutinized the difference between participatory performance-as-research and as pedagogy. In the area of work with children and teenagers, 'performance' and 'participation' as well as 'learning through research' are increasingly promoted by social forces which combine their idea of a participatory culture with a rather conservative notion of what that culture entails.[14] Moreover, these forces tend to reduce performance work and 'learning through research' to the transmission of economically useful 'key-skills' such as self-responsibility, creativity, and flexibility.

In opposition to such reductionism, performance research will have a much greater potential if the various transitions between performance and research are allowed to evolve. This may lead to a different understanding of research, which regards it no longer as the privilege of the specialist fields of science, theory, and art. Instead, the role of these fields would be to help organize and make visible the collective research that is undertaken by everybody, every day, making use of a wide range of procedures and integrating all forms of knowledge. In this sense, performance research would be linked to a certain egalitarian politics in our knowledge society. Yes, if such transitions are not developed, 'performance' may simply come to stand for 'output,' and 'research' for what economists do when they analyze markets. What would a 'performance research' of the latter kind make of the model of performance research I have outlined above? It is not coincidental that the future of work is a topic that has of late featured prominently in the research on performance as well as in performance-as-research. What is at stake here is the thesis that the performer with his or her adaptability provides the template for the ideal worker of the future. But if the fields of research are to blend into each other as I have claimed, making scholars and artists interchangeable, wouldn't that also prove this thesis right in the end?

Notes

1. In his study on *Postdramatic Theater* (2006) Hans-Thies Lehmann, in contrast, differentiates between theatre and performance art. Countering the idea of a

spread of a sense of 'staging' through other art forms, he describes the movement of theatre toward performance art as being motivated by an experience of the real.

2. http://www.sfb-performativ.de.

3. For an overview refer to Wirth, 2002; Fischer-Lichte and Kolesch, 1998; Fischer-Lichte and Wulf, 2001, 2004.

4. Regarding the philosophy of language, in the 1980s German philosophers Jürgen Habermas and Karl-Otto Apel developed the 'avoidance of the performative contradiction' as the core of their theory of communicative action – a concept that has been criticized profoundly by French postmodern theory (see Habermas, 1984; and Lyotard, 1984).

5. The mutuality of studies of science and performance research can easily be grasped when considering how, for example, historians of science have rebuilt historical experimental appliances rather than accepting textual evidence for experimental outcomes (Heering et al., 2000).

6. Somewhere along these lines I would locate my own research project, which is concerned with the lecture as performance or the lecture-performance. Its intention is to explore how the production and the presentation of knowledge are linked within the performative dimension of the lecture. Whilst I investigate this link primarily in the context of historical case studies, I also use the lecture performance as an experimental arrangement to analyze what in recent years has amounted to an explosion of the 'lecture performance' as a new hybrid format of presentation (Peters, 2005).

7. Thus, the academic position of a performance group like Theater der Versammlung (lit: Theatre of Assembly) is relatively unique: it is funded by the University of Bremen, where it has the mission to intervene regularly into the everyday of academic life and bring its performative dimensions to everybody's consciousness. (http://www.tdv.uni-bremen.de).

8. http://www.unfriendly-takeover.de.

9. Both summer academy projects are staged at performance venues, including, in Hurtzig's case, the Hebbel am Ufer HAU theatre, today one of the most important venues for performance in Germany.

10. Jörn J. Burmester, Florian Feigl, Otmar Wagner, Dariusz Kostyra and Christopher Hewitt are some of the artists involved (http://www.performerstammtisch.de).

11. Undergraduate programs in cultural studies (with theatre and performance studies as one focus among others), which combine theory and practice, now exist at the universities of Hildesheim and Lüneburg. In addition, some more traditional departments for theatre studies employ lecturers with special responsibility for conveying practical experiences to the students. For information on graduate programs in dance-focused performance studies, see the next section.

12. The participants' list on the congress's website (www.tanzkongress.de) reads like the 'who's who' of the German and international dance scene.

13. For the development of this kind of performance the work of Christoph Schlingensief, in particular his *Bahnhofsmission* (1997) and the election project *Chance 2000 – Wähle Dich Selbst* (*Vote for yourself*, 1998), has been highly influential.

14. An alternative practice of performative research with children is provided by the Forschungstheater ('research theatre') at the Fundus Theater in Hamburg

(http://www.fundus-theater.de), which has staged projects such as *Forschen für Anfänger* ('Research for Beginners'), a series of performative presentations by experts for children.

References

Blisset, L. and S. Brünzel. *Handbuch der Kommunikationsguerilla* (Berlin: Assoziation a, 2001)

Bormann, H.-F. 'Der unheimliche Beobachter. Chris Burden, 1975: Performance als Dokument,' in *Wahrnehmung und Medialität / Theatralität 3*, ed. E. Fischer-Lichte et al. (Tübingen and Basel: Francke, 2001), pp. 403–19.

Bormann, H.-F. and G. Brandstetter. 'An der Schwelle. Performance als Forschungslabor,' in *Schreiben auf Wasser. Performative Verfahren in Kunst, Wissenschaft und Bildung*, ed. H. Seitz (Essen: Klartext, 1999), pp. 45–56.

Brandstetter, G. 'Aufführung und Aufzeichnung–Kunst der Wissenschaft?,' in *Kunst der Aufführung–Aufführung der Kunst*, ed. E. Fischer-Lichte et al. (Berlin: Theater der Zeit, 2004), pp. 40–50.

Brandstetter, G. and S. Peters. (eds) *de figura. Rhetorik, Bewegung, Gestalt* (München: Fink, 2002).

Deleuze, G. and F. Guattari. *A Thousand Plateaus*, trans. B. Massumi (London: Athlone Press, 1988).

Duden, Rechtschreibung der deutschen Sprache, 21. Auflage (21st edn) (Mannheim, Leipzig, Wien, and Zürich: Bibliographisches Institut, 1996).

Felfe, R. 'Schauplätze des Wissens. Rückblick auf zwei Tagungen in Berlin,' *kunsttexte.de*, 3 (2002).

Fischer-Lichte, E. *Ästhetik des Performativen* (Frankfurt/M.: Edition Suhrkamp, 2004).

Fischer-Lichte, E. and D. Kolesch. (eds) *Paragrana. Internationale Zeitschrift für Historische Anthropologie*, 7 ('Kulturen des Performativen') (1998).

Fischer-Lichte, E. and J. Roselt. 'Attraktion des Augenblicks – Aufführung, Performance, Performativ und Performativität als theaterwissenschaftliche Begriffe,' *Paragrana*, 10 ('Theorien des Performativen') (2001) 237–54.

Fischer-Lichte, E. and C. Wulf. (eds) *Paragrana. Internationale Zeitschrift für Historische Anthropologie*, 10 ('Theorien des Performativen') (2001).

——. (eds) *Paragrana. Internationale Zeitschrift für Historische Anthropologie*, 13 ('Praktiken des Performativen') (2004).

Habermas, J. *The Theory of Communicative Action*, trans. T. McCarthy (Cambridge: Polity, 1984).

Hamacher, W. 'Afformative, Strike: Benjamin's "Critique of Violence",' trans. D. Hollander, in *Walter Benjamin's Philosophy: Destruction and Experience*, ed. A. Benjamin and P. Osborne (London and New York: Routledge, 1994), pp. 110–38.

Heering, P. et al. *Im Labor der Physikgeschichte: zur Untersuchung historischer Experimentalpraxis* (Oldenburg: Bibliotheks- und Informationssystem der Universität Oldenburg, 2000).

Institute for Applied Theatre Studies, 'Institute – Concept,' *Institute for Applied Theatre Studies (Institut für Angewandte Theaterwissenschaften) – Justus-Liebig-Universität Giessen website*, http://www.uni-giessen.de/theater/en/institute (accessed 13 December 2008).

Krämer, S. 'Sprache – Stimme – Schrift: Sieben Thesen über Performativität als Medialität,' *Paragrana*, 7 ('Kulturen des Performativen') (1998) 33–59.

Latour, B. and P. Weibel. *Making Things Public: Atmospheres of Democracy* (Cambridge, MA, London, and Karlsruhe: ZKM & MIT Press, 2005).

Lehmann, H.-T. *Postdramatic Theater*, trans. and intr. K. Jürs-Munby (London and New York: Routledge, 2006).

Lyotard, J. F. *The Postmodern Condition – A Report on Knowledge*, trans. G. Bennington and B. Massumi (Manchester: Manchester University Press, 1984).

Matzke, A. *Testen, Spielen, Tricksen, Scheitern. Formen szenischer Selbstinszenierung im zeitgenössischen Theater* (Hildesheim, Zürich, and New York: Olms, 2005).

Mersch, D. *Ereignis und Aura. Untersuchungen zu einer Ästhetik des Performativen* (Frankfurt/M.: Suhrkamp, 2002).

Peters, S. 'Sagen und Zeigen – der Vortrag als Performance,' in *Performance. Positionen zur zeitgenössischen szenischen Kunst*, ed. G. Klein and W. Sting (Bielefeld: Transcript, 2005).

Peters, S. and M. J. Schäfer. (eds) *Intellektuelle Anschauung – Figurationen von Evidenz zwischen Kunst und Wissen* (Bielefeld: transcript, 2006).

Schechner, R. *Performance Studies – An Introduction* (London and New York: Routledge, 2002).

Schramm, H. et al. (eds) *Spektakuläre Experimente. Praktiken der Evidenzproduktion im 17. Jahrhundert* (Berlin and New York: Walter de Gruyter, 2006).

Wirth, U. (ed.) *Performanz. Zwischen Sprachphilosophie und Kulturwissenschaften* (Frankfurt/M.: Suhrkamp, 2002).

10
Translate, or Else: Marking the Glocal Troubles of Performance Research in Croatia

Lada Čale Feldman and Marin Blažević

How to challenge the disciplinary constraints imposed by global theories of the ever-widening field of performance, beginning with the linguistic demand to translate – terms, theories, aesthetics, traditions, texts, contexts, performances – or else ... fall out of both the system of academic communication and the visible arena of aesthetic performative practices?

The demand to translate partly results from the peripheral position of the Croatian humanistic and social sciences in general, whose carriers feel compelled to assume the role of translators, consumers, and transmitters – perhaps also deformers – rather than performers and creators of theoretical insights, those who absorb and adapt theoretical influences from different disciplinary sources. In literary criticism, for instance, this has meant a *mélange* of approaches from Russian formalism, Anglo-American New Criticism, and Italian semiotics to French poststructuralism and German systems theory, sometimes successfully reformulated or even queered for the purposes of addressing specific issues generated by the local context and local knowledge. According to the Croatian literary and cultural theorist Vladimir Biti (2002), an aesthetic ideology and a belief in the centrality of literary texts governed the intense absorption and re-elaboration of critical paradigms in the founding years of modern Croatian literary criticism after the 1950s – even when such paradigms were accommodated within the ruling Marxist version of historical process.

This textual aesthetic had its counterpart in the ideology that ascribed centrality to great dramatic works during the artistically richest period of the socialist era, the late 1960s to the early 1980s – an ideology that is perhaps best exemplified by the establishment of national theatres and by the special state protection of the Dubrovnik summer festival. Serving the official discourse, which accorded to the arts in general the representative function of a progressively oriented society, this crypto-aestheticism at the same time perceived itself as a way to preserve national tradition and historical identity. It thus subtly resisted the imagined and real pressures of the once unitarian and then, after the turbulent 1970s, gradually less centralized and more tolerant communist regime. It constitutes a paradox resulting from several crucial cultural events in the 1950s, which explicitly announced the break with socialist-realist aesthetics that still, presumably, continued to hover in the party-programmed minds of bureaucratic custodians and financers of culture.[1] Hence we discern an impulse in both mainstream artistic practice and the academy to protect their fields of influence from any redefinition that would threaten their borders, a redefinition clearly implied by the arrival of 'performance' and its studies.

The emergence of the term

Any survey of performance research in Croatia has to start by recognizing a troubled disciplinary field without an institutionally established academic influence. The translation of the very terms 'performance,' 'performative,' and 'performativity' gets right to the heart of the trouble. The disciplinary traditions ruling over the reception, translation, and implementation of these terms were various and sometimes divergent, if not in direct conflict. What united them, however, is their lack of methodological and terminological autonomy, and their constant reliance on imported theoretical currents, whose selection and divulgation often directly depended on a rather crude pragmatic fact, namely the knowledge of foreign languages. Characteristic of major Croatian scholars, to varying degree, has been an eagerness to learn and spread the news that was emerging in different national theoretical contexts.

Thus, for several decades after the Second World War, German historically oriented traditions of *Literatur-* and *Theaterwissenschaft* and *Volkskunde*, Russian formalism, and Prague structuralism were highly influential in the constitution of a Croatian terminological apparatus for literary theory, folklore studies, and theatre studies. These were the three major disciplinary fields that would eventually welcome performance theory as an opportunity to reformulate their objects of research, approaches to it, and even their position with respect to the dominant ideological framework that (presumably or actually) controlled the relevance, the scope, and the outcome of their findings. In literary theory, the 1970s brought the first translations of French structuralism, narratology, and the Tartu school; the 1980s brought Eco's semiotics, and theatre studies added to it fragments of Kowzan's, Helbo's, and Ubersfeld's semiology of theatre. The late 1980s saw the translation of excerpts from Schechner's *Performance Theory*, and, in 1990, a rather belated translation of Turner's *From Ritual to Theatre*.

In the 1990s it might have seemed as if these translations had not made any substantial impact on Croatian scholars, for whom, it was still being said, 'early' semiotics was the latest news (Zlatar, 1996: 222). This was not, however, entirely true. By 1996, the parallel methodological protocols of semiotics and performance analysis were already taking hold due to a shift in generations: younger scholars felt the need to confront a widespread anti-theoretical stance that dominated both literary and theatre studies. And yet, a strict division still prevailed in theatre studies: both its main academic institutions – the Academy for Dramatic Art and the Institute for the History of Literature, Theatre and Music within the Croatian Academy of Arts and Sciences – divided their labor between the philological and comparative study of drama on the one hand, and the historical study of material conditions of theatrical production on the other.

It is therefore perhaps not surprising that it is a folklore researcher, Maja Bošković-Stulli, who gets the credit for first introducing and translating the term 'performance.' It was she who, having participated at the Helsinki conference on folk narrative research held in 1974, first came in touch with the approach presented by American folklorists Dan Ben Amos and Alan Dundes. They proposed to study folklore genres in the context of their production and reception, dealing with them as forms of live oral communication within small groups. As opposed to the inherited philological categories and modernist dichotomies that had characterized Croatian folklore research from its Herderian beginnings in the nineteenth century – with its pervasive rhetoric of collecting folk 'treasure' and its methods focusing on the authenticity, tradition, and the variability and aesthetic nature of folklore – American folklore researchers emphasized, according to Bošković-Stulli, 'the folklore event itself' (1978: 14).

Bošković-Stulli was not herself entirely enthusiastic about the wholesale implementation of the American methodological framework, since it appeared to neglect the literary qualities of folk poetry as a Jakobsonian 'aesthetic message,' 'objectifying itself' intermittently in repeated communications under different social conditions. She pointed to the pitfalls of the exclusive conception of folklore as a communicational act (*čin*), event (*događaj*), or performance. In her text, 'performance' alternated between a translation into Croatian (*izvedba*) and a 'Croatized' version (*performancija*). For Bošković-Stulli, the emphasis on performance was a logical reaction to the conception of folklore as a fixed text, but it presented several insurmountable challenges: how was it possible to grasp folklore communication 'on the spot,' since it often arose spontaneously? How could one rely on one's own 'subjective perceptions' and 'limited memory,' since there was no way to depend on 'precise documentation'? And how to account for the process of transmission, implied in the etymological root of the very term 'tradition'?

To lose this historical perspective for the sake of the study of the 'event itself,' Bošković-Stulli concluded, would severely damage not only the understanding of folklore as a long-standing historical communicational process, but also the understanding of any individual performance. If it were to become the focus of research, it would figure as part of an interest in human behavior in general, but it would not be a contribution to the study of 'literary aspects' of 'folklore expression.' Nonetheless, Bošković-Stulli endorsed Dundes's terminological division of folklore into 'text,' 'texture,' and 'context,' since this triad enabled the researcher to maintain the distinction between 'text' and 'texture' on the one hand, and the 'dramatization of the texture' and 'context' on the other. And yet, this solution flagrantly contradicted the idea that the folklore text does not exist 'outside' of its individual utterances – at least not without the help of a zealous collector and transcriber applying a methodology elaborated in the study of written texts.

Despite her critical stance, Bošković-Stulli was the first to adopt and translate the term 'performance' by using the Croatian word *izvedba*. This is a term used in theatre studies to mean 'enactment.' Often, unfortunately, this usage is connected with precisely the understanding of theatre which the concept of 'performance' has tried to subvert, that is, theatrical performance as first and foremost 'the enactment' (*izvedba*) of a pre-existent play.

'The flicker of tragic sight'

The local emergence of the term 'performance' within folklore studies may not be surprising today. But it is startling that some voices are still trying to contest attempts to engage with the use of the term in the theoretical writings of director Branko Gavella, the most prominent representative of Croatian Modernist theatre. Although there have been several attempts to subvert and alter the dominant perception of Gavella as a sacrosanct patron of literary-dramatic, verbal, and mimetic theatre, his supposed 'advocacy of literature in the theatre' (Gavella) still reverberates in the airless spaces of the dominant institutions of Croatian theatre, above all in the Croatian National Theatre and the Academy of Drama Art, the only Croatian institution of higher education in theatre.

On closer inspection, Gavella's interdisciplinary theoretical writings on performance dissolve the central tenets of his famous formulation cited above and reveal that, naturally, the correlations between the literary/dramatic text and its theatrical representation cannot be boiled down to the function of 'advocacy.' Here, however, we shall not explore Gavella's analysis of this 'dramaturgical problem.'[2] Instead, we will focus our attention on two other implications of Gavella's theory, even though this means excerpting only a small sample from the impressive range of his theoretical discourse.

Gavella analyzed developments in the style of Croatian theatre since its so-called 'modern revival' in the mid-nineteenth century, and in the process he located two long-lasting struggles that joined literature and theatre in a 'capillary correlation': first the struggle for the right to a national language[3] and second the struggle to create a standardized version of that national language that would overcome 'folklore speech-models' and 'patterns of dialects' (Gavella, 1968: 23). From the mid-nineteenth century to the early twentieth century, literature gradually began to 'impose its own standards' on theatrical production. At the same time, entangled in specific local political situations, theatre served propagandistic and didactic causes. Therefore the modern revival of Croatian theatre took place in a 'critical phase,' when 'literary production, under pressure of its ideological overdetermination and its nationally-representative aspirations, was looking for the widest possible range of its words, and when the rapid development of the modern concept of the state attempted to subject to its interests all public positions, from which the word could be aimed towards the definition and "re-education" of the listeners' (Gavella, 1953: 12). If the convergence of theatre, literature, and even oppositional political agendas was initially emancipatory, over time this union turned into an obstacle in the development of theatre: the performance of actors, formerly 'language pioneers,' became not just 'relatively poor in gesture and facial expression' but also reflective of the actor's 'bureaucratical affection.' Curiously, 100 years later, at the beginning of the twenty-first century, performing arts and performance research in Croatia have faced the question of language once again, but this time in glocal relations. Should one perform in Croatian, or in English?

The second implication of Gavella's theory brings us into the 'proximity of performance.' Although it might be too ambitious to declare that Gavella's theory anticipated current assertions that performance art was born 'within and against theatrical form' (Heathfield, 2004: 8), it should be noted that Gavella drew attention to a performative fissure even within such a classical theatrical form as tragedy. Inspired by the 'artistic peculiarity' of Lawrence Olivier's acting as Titus Andronicus and Macbeth,[4] Gavella first claimed that a 'general sense of the tragic [...] involves revealing the depth of experience every human individual carries by virtue of his predicament in the human community [...] beneath his public face,' and he suggested that 'the essence of the tragic lies in removing this mask, in the fight for this mask, in the gestures of this fight, [...] and in the final outcome of this fight, when in its flickering flames the naked face of man is shown – naked because it removes all screens from piercing view into the essence of man's inner being' (Gavella, 1970: 8; translation Blažević). The words 'mask' and 'nakedness' are not just frozen into metaphors when they are applied to the embodied performance of the actor. Gavella thus reveals that, at heart, the tragic 'is not a literary but essentially a theatre-stage category.' And, curiously, it is through his understanding of the tragic that outlines of the phenomena and experience of performance appear. Gavella argues that 'only the actor's experience – only the dependence of this experience on the immediate insight into the spectators' experience – can reveal the meaning of removing the mask, present the horror of the sight when the pale flame flickers through the living, naked eye and the tone of human suffering quivers on the living mouth' (Gavella, 2001: 6). Giving prominence to the immediacy and experiential impact of acting, its distinctive features springing from the fact and the actions of the present living body of the actor, Gavella suggests that what Heathfield calls 'the drive to the live,' which 'has long been the critical concern of performance and Live Art' (Heathfield, 2004: 7), has been a constitutive element of theatrical enactment even longer, more or less foregrounded (or 'masked') in various historical aesthetics.

Furthermore, Gavella anticipates the ontological definition of (here also theatrical) performance as it was conceptualized by Peggy Phelan in *Unmarked*. According to Phelan, as we know: 'Performance's only life is in the present. Performance cannot be saved, recorded, documented, or otherwise participate in the circulation of representations *of* representations [...]' (Phelan, 1993: 146). Or, in the even more emphatic words of the Croatian director and theorist fascinated by Olivier's acting performance: 'These are all phenomena of observing and listening, but no hue, no palette can transmit this observing, for the palette is always static and bears the trace of the hand and the brush, while the flicker of tragic sight comes from depths touched by no outside hand, from occurrences that can be recorded by no musical instrument, save the fundamental instrument that human experience fashions in profound need of expression' (Gavella, 2001: 6).

If a certain anti-representational potential is always already immanent to theatre in the very actor's performance, it might be that the antagonistic 'drive,' which according to Heathfield leads to the birth of performance, is directed more against the norms that the institution of theatre is imposing on the form of theatre than against the presumed theatrical form as such. Or, from a different perspective, it also might be claimed that it is precisely the institution of performance – or live – art that reduces the theatrical form, or rather the multiplicity of its forms, to only a few norms (impersonation, narrative, mimesis, identification), in order to secure its own symbolic (counter-)capital on the cultural market. And yet, these relational models cannot provide a persuasive account of the local situation since, until recently in Croatia, it was impossible to imagine – let alone strike – a balance of institutional interests, support, power, positions, and perspectives between dramatic theatre on the one hand, and performance or live art on the other. The friction between theatre and performance, already identified by Gavella in the actor's 'removal of the mask,' was lessened by the overriding representational function that dominated the performing arts – by the end of the 1960s reduced to only three forms: dramatic theatre, opera, and ballet. Arguably, they disregarded this friction in the interests of an undisturbed representation of 'great' literature, national or simply 'high' culture, and, more or less explicitly, hegemonic political ideology. When a new awareness regarding the aesthetic, conceptual, and political challenges that could come from performance liberated the performing arts from the dominance of the representational function, and when the performative force and the affective potential of the living body infected the theatrical enactments of several independent groups of artists and gave birth to happening and performance art on the local scene, the institutions of theatrical production and theatre studies excluded these new possibilities – an exclusion that lasted until the late 1980s. Removal from theatrical institutions hurt the new theatre companies (also called 'alternative,' 'neo-avant-garde,' or 'youth' theatre), while it made happening and performance art invisible within the domain and the institution of theatre in general.

From performance to theatrability

Both the philological approach to drama as a literary text and the urge of folklore researchers to preserve oral literature in the form of written, fixed texts, to be studied for stylistic features, contributed to the idea that folklore researchers should collect and notate 'folk drama.' According to the classical division between epic, lyric, and dramatic works, which was being applied to folklore genres in the 1960s, drama was the missing third term. Ivan Lozica, a newcomer to this field who was working in close cooperation with Nikola Bonifačić Rožin, the foremost proponent of the existence of 'folk drama,' argued strongly against the uncritical transfer of such literary classifications, thereby following the avenues opened up by Bošković-Stulli and her endeavors to negotiate between Prague structuralism and recent American contextual folklore studies. However, Lozica's interest in 'representational folk art' – the representational and dramaturgical aspects of 'customs, games, dances, and rituals taking place within the context of everyday life in both rural and urban settings' (Lozica, 1990: 8; translation Čale Feldman) – crossed over into what had until then been called the ethnological 'context.' It also called for new terms and concepts, no longer provided by the notation of folk drama dialogues, since the author's turn from the outdated notion of 'dramatic conflict' – a nineteenth-century term thoroughly inappropriate for folklore dialogical exchanges – to the notion of 'representation' – the 'showing of an imagined reality' through costume, masking, voice, gestures, movement, dance, and the often improvised interaction of performers and audience – doubly threatened the disciplinary field, which could anticipate its losses in confrontation with the aesthetic norms of theatre studies on the one side, or its uneasy conflation with ethnological everyday 'banality' on the other.

Following Bogatyrev's and Mukařovský's dynamic conceptions of the changing functions of a given sign-structure, Dundes's distinctions between 'text,' 'texture,' and 'context,' and Hymes's differentiation among 'behavior,' 'conduct,' and 'performance,' Lozica proposed his own gradation of what he called the 'representational continuum' – from individual 'theatrical behavior' to 'theatrability' to 'theatre' (Losica, 1990: 78–81; translation Čale Feldman). These terms reflected the progression of the role and impact of the aesthetic function with respect to possible other functions – magical, socio-critical, and so on – in representational events. His main thesis was that theatre was not 'born' out of (folk) ritual once, but that it undergoes continuous and repeated 'births' out of numerous behavioral practices which, in certain circumstances, can acquire aesthetic and entertaining functions and intensify these at the expense of the previously dominant ones.

This thesis, of course, put a lot of weight on the 'theatrical frame,' a conceptualization borrowed from Goffman, which ultimately allowed the factors governing the establishment of the frame – the onlookers and their levels of understanding and classifying a given event – to mark the fictional status and the aesthetic impact of the event, rather than the 'texture' of 'the event itself.' Very much in the vein of Schechner's all-encompassing performance 'web,' Lozica's representational continuum put the notion of a 'proper' theatrical performance squarely in its transitory place, proclaiming it the outcome of just one contingent type of theatrical practice, whose institutionalization and cultural sanctification was bourgeois in character and resulted from a sedimentation of numerous historical, political, and socio-cultural factors.

Although Lozica's own research moved and continues to move primarily within the limits of folklore 'theatrability,' legitimating the hybrid disciplinary field of 'ethno-theatre studies,' as he decided to call it, it is important to note his sporadic use of the 'Croatized' designation 'performative folklore genres' (*performativni oblici folklora*) for the object of his interest. This term, along with his suggestion that performance analysis should, in principle, include a broad range of performative practices – from children's games to circus to everyday performativity – largely justified the title of his most important study, *Outside of Theatre* (1990), which marked a crucial methodological and aesthetic 'breakthrough into performance.'

Struggling for a change

In 1956 Gavella published 'Theatre Impressions from London' (see Gavella, 2001), in which he inverted the hierarchy of the constituents of the theatrical event and foregrounded the primacy of the (organic) experience of live performance. In May 1963 John Cage, while visiting Zagreb to participate in the Music Biennale, an international festival of contemporary music, got together with members of the neo-avant-garde, the 'undefined and undeterminable' Group Gorgona (active 1959–66), whose 'aspirations,' according to one of its authors, the painter Josip Vaništa, 'were directed towards an extra-aesthetic reality' ('Gorgona sometimes did nothing, it just lived'). And just a few years later, in 1967, Tomislav Gotovac, one of the founders and a key-figure of performance art in Croatia, performed *Happ Our-Happening*, in which he demolished furniture and threw trash into the audience. At the time, it was inconceivable that any correlation could be established between the most daring insights of Gavella's theory and a new form of performing that brought the contingency and liveness of performance to the fore.

In the meantime, as is well known, a new research paradigm – performance studies – has evolved in a global context with an ambition to span the full range of cultural performance, including both dramatic theatre and a great variety of 'nonmatrixed' performing (Kirby, 1995 a,b). Locally in Croatia, during the 1970s and 1980s, performance art gained the status of an approved, albeit liminal, even 'misfit,'[5] manifestation of visual, (neo)conceptual or so-called contemporary art. However, it was not until the late 1980s that the newly founded international Festival of New Theatre, Eurokaz, attempted to link the whole gamut of artistic practices in-between performance art and theatre by using the vague idea of the 'new theatre.' And it was not until the mid-1990s that the magazine *Frakcija* (Fraction) introduced a comprehensive generic term – 'performing arts' (*'izvedbene umjetnosti'*) – to bring theatre and performance art closer to each other within the sphere of theory and criticism.

Still, it would be wrong to presume that this drawn-out scenario was the unfortunate outcome of some local cultural restraints, even of a kind of retardation, or solely of the exclusivity of theatrical institutions. Despite all of the troubles with production and reception, the process of questioning, deconstructing, even of deforming theatrical form, initiated in the 1960s, would in the following decades come to develop various hybrid formations and strategies of (transgressive, resisting, radical, counter-)performance *outside* of the dramatic, academic, national theatre, yet *within* the extended frame, form or, if you like, paradigm of (postdramatic) theatre. It appears that the persistence of the collusions and collisions between theatrical representation and performative effect, between (live) theatre and performance (art), was a challenge that local representatives of 'alternative,' 'new' or (as we will term it from now on) 'performance-theatre' were not willing to throw to the winds. Moreover, they did not completely discard the constitutive elements of drama – dramatic situation and action, characters and conflict, fiction and narrative, mimesis and illusion – although all those elements had been undermined, fragmented, fractured, disjointed, discorded, disarranged, and submerged by a performance flow.

Performance-theatre in Croatia was evidently inclined to place itself against the institution(s) of dramatic theatre, as well as against the repressive authoritarian regimes which, until recently, had been restricting cultural, social, and political life throughout the turbulent history of Croatia inside and outside of Yugoslavia. The proponents of performance-theatre did so by stressing contingency, liminality, the transitory nature of performance, its primary dependence on affective and defective potential, and also by turning to techniques of the body, rather than to the semantic purposefulness of the word and the totality of the represented (fictional) world. They did so, too, by (self-referentially) rethinking the whole complex of the interdependent and interwoven factors, functions, and effects of (theatrical) representation and performance, from the strata of identity of the actor-performer to the politics of maintaining or breaking theatrical (or political) illusions. They resisted institutional and political norms by experimenting with various tactics to elicit the physical and verbal participation of the spectators in a democratized theatrical space. And, most of all, they explored de-hierarchized models of production and (self-)organization of (at least notionally) independent artistic as well as social micro-communities. The question arises, however, why and how theatrical form and elements of drama were nevertheless incorporated into such a strategy of counter-performing.

Arguably, the obstinate persistence of the form and elements of drama and theatre might have had to do with a certain lack of an avant-garde tradition in performance. That is, the origins and development of performance art could not have been traced all the way back to the historical avant-gardes since, aside from a few exceptions, Croatian theatre practice had no significant avant-garde achievements, at least none that could live up to the challenges offered by avant-garde dramatic literature. But the second main reason for the persistent channeling of performance towards the seemingly obsolete functions of representation was the inclination of performance-theatre to take part in socio-aesthetic drama. What we have in mind here is a struggle for both aesthetic and socio-political change, even radical cultural transformation (especially in the work of Kugla theatre and Damir Bartol Indoš). In order to stage its antagonistic position, to perform the 'breach' and cause the 'crisis,' to demonstrate possibilities and ways of change, performance-theatre was strategically keeping alive the deconstructed elements and forms of the twin protagonists of the socio-aesthetic status quo – institutionalized dramatic theatre anchored in realism on the one hand, and on the other hand the highly theatricalized reality of 'real socialism' or post-communist hegemonic nationalism, which more or less efficaciously suppressed and marginalized any still daring enough to attempt to bring about a potential conflict.

Thus it does not come as a surprise that such counter-politics of performance produced reactions on the side of the ruling theatrical institutions, their representatives and beneficiaries, creators and supervisors in the sphere of cultural politics. They applied – and still apply – various strategies to violate 'performing rights', ranging from withholding financial support to launching discursive offensives in the media. Although it would be unjustified to claim that the performance-theatre was systematically and continuously suffocated during the whole period from the late 1960s to the late 1990s, from today's perspective it is clear that the establishment was rather inarticulate when provoked to define the limits of acceptable activity of its antagonists, but it was tenacious when forced to defend those borders. Despite several exceptional cases when those limits were breached – especially in the second half of the 1980s, just at the threshold of transition – the cultural and political position of performance-theatre did not significantly change until the end of the 1990s.

On the one hand, the ultra-conservative policy and politics of Franjo Tudjman were already dying out at that time, which resulted in a gradual and relative liberalization of cultural life. On the other hand, the overall development of interdisciplinary performance research not only engaged the local fields of literary theory, ethnography, and theatre studies, but also led to the launching of a performing arts journal (*Frakcija*) that has shifted the focus of performance research to the concrete and topical production of performance-theatre,[6] and that has also assumed the responsibility for the translation of performance theory and terms from a global to the local language and context, and – since 1999, when it started to appear as a bi-lingual edition (Croatian and English, sometimes even only in English) – vice versa. It was mainly *Frakcija* that finally provided a discursive space and support for a kind of performing (and) theory which is above all preoccupied with the bare fact and volatility of the flow of performance, then with its (self-)reflexive twisting, its 'social efficacy,' and its general cultural conditioning.[7] Paradoxically, despite its efforts to submit to the double demands of local-global and global-local translation, it now seems that, just as *Frakcija* has achieved a certain level of international visibility, it is gradually losing its own local readers, and therefore its local 'social efficacy.'

In this respect, *Frakcija* has to deal with a similar performative blackmail (perform/translate or else ...) to the one faced by those performance-theatre companies and authors that – in order to become internationally relevant (and marketable) – display an inclination to attune both the performative strategies and the referential horizons of their performances to trouble-free translocal recognizability and translatability, even introducing English as the second 'official' language of enactment. Captured in these transitional/translational paradoxes, *Frakcija* currently seems to be looking for an urgent response to the following challenge: how far can and should it move away from its *intra*cultural tensions, its local context, knowledge, concerns, even language, and – dare we say – tradition, in order to increase the efficacy of its 'global' performance? And at what cost?

The ethics of the performative

Performance, obviously, has dissolved boundaries: not only those of the entity under scrutiny, but also those of the scrutiny itself. The notion of performance entered the literary-dramatic field at a moment when there was a turn away from a philological and phenomenological concentration on the text, towards the transdisciplinary enterprise of exploding the integrity of 'high' literature as the unquestioned 'object' of study. The suggestion that high literature had been defined by unacknowledged hierarchies of value issuing from class, gender, and national biases had a clear ethico-political resonance in the attack on the Croatian literary-critical academy by the Croatian theorist Vladimir Bit, whom we mentioned earlier, especially in the context of a renewed ideology of nationhood in times of transition, with literature as its consecrated expression. Biti's 700-page *Concepts in Literary and Cultural Criticism* (Biti, 2002)[8] was the summation of his extensive work on the main schools of Anglo-American, French, and German thought, embracing philosophers, linguists, literary and cultural theorists. In the early 1990s, this engagement took the form of a forum, in which young scholars led by Biti regularly debated issues involved in the localizing of globally influential critical paradigms.

Among the 350 ruling concepts of contemporary theory that Biti chose to discuss, 'performance' figures in its Croatian translation (*'izvedba'*), while the entry on the untranslated, only slightly Croatized 'performative' (*'performativ'*) is left blank, merely pointing to entries on the related concepts of 'performance' (*'izvedba'*), 'act' (*'čin'*), and 'illocution' (*'ilokucija'*). On closer inspection, however, Biti's entry on 'performance' reveals a certain set of reservations about the non-linguistic impact and destiny of the term. He acknowledges the importance of performance analysis in replacing the close reading of dramatic texts and explicitly states that 'the theatrical connotations of the term are not accidental,' issuing as they do from 'the tradition of understanding language through the analytical philosophy of L. Wittgenstein or the American sociologist E. Goffman' (Biti, 2002: 237; translation Čale Feldman) and coming from the pragmatics of discourse by T. van Dijk, as well as from the semantic blending of 'stage and sexual performance' in Felman's interpretation of Austin. But Biti has strong suspicions about the potential workings of the term whenever it attempts to embrace conceptually what he once praised as the final breakthrough of literary theory into the 'overall gamut of human thought, activity and behavior' (Biti, 1995: 110–11), suspicions that concern particularly the latter two aspects. Thus, the last paragraph of the entry does not hesitate to denounce 'certain difficulties of performativism (*performativizam*)' both from a 'research' and a 'practical perspective,' since 'the sources that testify to performance are often found to be difficult to trace, lacking or limited in usefulness,' while, on the other hand, no 'interpretative whole' (Biti, 2002: 239; translation Čale Feldman), according to the author, could ever manage to unite the entire constellation of events described in the well-known definition offered by Schechner in his *Performance Theory*. We might ask, however, whether this reservation involves forgetting that such an interpretative whole could never be reached when dealing with a literary text either?

Despite Biti's concerns, the relevance of Austin's 'performative' for literary analysis,[9] demonstrated in the Croatian translation of the widely appraised defense of Austin in Shoshana Felman's *Scandale du corps parlant* (1993), has already produced an impact on the younger scholarly elite, primarily literary critics Tatjana Jukić and Morana Čale. Their continuing interest in using the performative for purposes of queering extant modes of Croatian *Literaturwissenschaft* (Jukić, 1999; Čale, 2001) not only introduced a new performative quality to criticism itself – abandoning the aim to fix the identity of The Text for the sake of its interpretive reinscription which would make manifest its disseminative field of endless productivity – but also inspired a project entitled *Easy Said, Easy Done: Ethics of the Performative*. It gathered researchers in literary, media, and cultural studies who were interested precisely in the traces of performative disturbance and the breaching of the borders of the linguistic code by its either undisclosed or unpredictable future modes of embodiment. They also implied an awareness of historicity and politics that the previous concentration on the rhetoric of discourse had seemed to keep at bay. The 1990s, we must remember, were marked by huge historical and political local trouble – the war in Croatia – and by the impact not only of various conflicting political discourses and media reports, but also of the uncertain role played by writers and intellectuals in either stirring up or calming the conflict. This made the 'ethics of the performative' an acute problem of daily experience. The collapse of the old state and the establishment of the new one, however, coincided also with a widening of analytical interest in 'public events' (Handelman, 1990) that happened outside of theatre: including anti-war protests, new public genres of electoral gatherings, ritual consecrations of new state-protocols, the inauguration of the first democratically elected Croatian President, and new celebrations such as the Day of Statehood, which have been analyzed as a performance strand of 'war ethnography' (Čale Feldman Prica and Senjković, 1993). This moment also opened up avenues for a performance-oriented political and feminist anthropology (Čale Feldman, 1995, 2001a, 2001b, 2002, 2005), the latter – together with its interest in aesthetic performances – having been particularly inspired by various discussions of Judith Butler's works.[10]

The ethics of performance

The overall concern of this chapter has been to present the local landscape of performance research in a kind of zigzagging of perspectives, from traditional theatre studies to folklore research, from Gavella's theory of acting to cultural theory, from the politics of space to editing and translation politics. But the local landscape of performance could never be adequately understood without an insight into the disruptive force of aesthetic performances. By way of a conclusion – but also a meta-explanatory frame for our chapter – we now turn from concepts and contexts to performative practice, which itself has provided implicit guidelines for the story we have just told.

Manifold cultural components of experience, particularly those pertaining to the bodily 'tissue' of performance, have been and still are explored through multi-layered events, meta-theatrical procedures and auto-performative intensities in aesthetic performances produced by various performance-theatre groups and artists in the past four decades. We shall highlight four topics which we read as characteristic of these aesthetic projects; these are counterparts to the main concerns of this chapter.[11]

1   The text in/of (theatrical) performance and the (theoretical, institutional, ideological) demand for the 'interpretative whole'

Going beyond the Croatian tradition of – even obsession with – the 'advocacy of literature in theatre,' the performance *Ormitha Macarounada* (1982) by Coccolemocco mechanically merged incongruent texts, namely Pinter's *The Dumb Waiter* and Gide's novel *The Counterfeiters*. In radical opposition to the theological machinery of dramatic theatre, the group provoked collisions between the texts and stretched the textual material to the point of either blocking the performance due to the excessiveness and redundancy of signs and senses, or liberating the new signifying material and freeing the affective potential of the performance of authorial pressure: '40% of the material of that kind of performance is completely senseless, you can discard it. [...] But there is also the other 40% of the meaning which confirms a flashing, estranging compatibility of the seemingly incompatible. The remaining 20% is in the service of a stately endless wasting of time [...]' (Brezovec, 1999: 154).

2   (The function of) theatre inside/outside (everyday, social, political) reality and the treatment of the 'frame'

Contrary to current practices of de-theatricalizing performance by incorporating real, everyday life/material into the referential horizon of (pre-)texts and actions occurring within its flow, Kugla theatre researched various modes of theatricalizing everyday life. But they did so in a way that was entirely different from the ways that political spectacles sustain their ideological illusions. By means of carnivalesque, oneiric, psychedelic, and estranged happenings, as well as images, compositions, processions, and environmental interventions, Kugla transformed spaces of everyday life into fantastic ambiences, landscapes of imagination, psycho-dramatic sites for experimenting with a range of freedoms – of form, style, genre, social relations and organizations, representational material and performative actions, and expression itself. The dream, hypnosis, hallucination, ritual surrealist play, circus or cabaret – these were the transitional states and situations of representational vagueness, into which Kugla transferred characters, performer-actors and spectators, thus creating a kind of subversive utopia.

3   The (actor's 'flickering') performance in-between (the body's) 'presentness' and 'representability'

Performances by Damir Bartol Indoš often explore the stamina of the body strained between symbolic charge and radicalized presence. On the one hand, eruptions of corporeality (convulsions, contortions, tremors, imploring, howling, groaning, screeching, yelling, panting, glowing, and sweating), even high-risk life-threatening acts, resist the compulsion to signify and confront the spectator with the shocking reality of the skin, bones, flesh, secretions, injuries, hematomas, the pain, the breath. This kind of performing body could ultimately not only release the perceptive and organic experiential potential from the chains of semiosis, but could also lead to the transgression of the representational frame as the first act in transforming oppressive socio-cultural roles, norms, and functions. On the other hand, spectacular symbolic actions – a series of referentially legible movements, sounds, images, even explicit messages – burst into the tempestuous flow of the performance and impose themselves as unavoidable challenges, appeals for the urgent, concrete – as Indoš would say – 'constructive activity': from counter-cultural action in an Artaudian vein to the revolt against all kinds of repression – social, political, religious, medical, technological, even biological.

4 (The ethics of) the politics and economics of translation (from language to the whole gamut of cultural performance)

Against the Wilsonian spectacle of 'a kind of universal history that appears as a multicultural, ethnological, archaeological *kaleidoscope*' (Lehmann, 2006: 79; original emphasis) – which Vanden Heuvel has criticized as an escapism that leads from the beauty of the images into a 'cultural kitsch' (1991: 177) – Brezovec produces performances particularly sensitive to the vibrations of a multi-ethnic environment, which refer to local cultural, historical, even urgent political themes, dramas, and traumas. In these performances, he pays special attention to a cultural *Gestus* of sorts, particularly with respect to (even traditional) style, body technique, and (para)linguistic features of acting. These are not taken primarily as means of representation and expression, let alone multi- or intercultural decoration. Instead, these elements are relevant as factors appearing within the cultural-anthropological and current socio-political polemics that theatrical performance stages in the flows of semiosis that it initiates. Being intraculturally committed and contextually overtaxed, most of Brezovec's performances are difficult to understand, let alone translate, outside of their local environment. That is what makes them firmly resistant to the hegemonic economy of the global international cultural market, whether of postdramatic theatre or performance art. Performance research including.

Čudna rabota da vazda vile od satira bježe. I nije čudo: pitomi s divjijemi ne imaju što činit. Ma gdje sam ja ovo? Jesam li ja Stijepo? Je li ovo naša kuća? [...] I Stijepo sam i satir sam [...].

(from *Skup* by Marin Držić)

Notes

1. For the bureaucratic intricacies governing the ex-Yugoslav system of financing culture, see Cvjetičanin and Katunarić, 1999, particularly pp. 19–23.
2. For a more detailed presentation of Gavella's theory, see Blažević, 2003a, 2005.

3. A little historical background on local language troubles: until 1847 the official language in the Croatian Parliament was Latin. After a short break during which Croatian was the official language, the German language reigned from 1854 until 1861, when the Emperor Franz Joseph agreed to introduce the vernacular. Until the mid-nineteenth century German, Latin, and Hungarian dominated the school-system.

4. Gavella saw the performances of *Titus Andronicus* (directed by Peter Brook) and *Macbeth* (directed by Glen Byam Shaw) during his visit to London in 1956. A fragment from 'Theatre Impressions from London,' from which we quoted above, was published in *Frakcija* (Gavella, 2001).

5. For broader information concerning 'misfit' performance artists see Milovac, 2002.

6. It was in *Frakcija* that performance analysis finally overcame the boundaries and generic conventions of theatre criticism and the hermeneutic approach to *performance text*. See Marjanić (2001) on *The Rocking* by D. B. Indoš; Pristaš (2000) on contemporary dance productions; Sajko (2001/2002) on *Oedipus* by Ivica Buljan; Vujanović (2001/2002) on *Terrible Fish* by Borut Šeparović and Performing Unit; Cvejić and Milohnic (2002/2003) on *Solo Me* by BADco; Blažević (2001) on Brezovec's *The Grand Master of All the Scoundrels*; Blažević (2003b) on Brezovec's *Kamov, Deathwrit & Moulin Rouge*, and *Heimspiel* by Bobo Jelčić and Nataša Rajković.

7. For the range of authors and themes represented in *Frakcija*, see the magazine or check: http://www.cdu.hr.

8. The first edition appeared as *Concepts in Literary Criticism* in 2000 and was translated into German in 2001.

9. As for the first translations of Austin, Strawson, and other representatives of pragmatic linguistics, parts of their work appeared in the late 1970s in the magazine *Dometi*, thanks to the endeavors of a philosopher, Nenad Miščević, himself author of a book entitled *Language as Doing*, which appeared in Slovenian in 1983. A part of his book, along with a chapter from Austin's *How to do Things with Words*, was translated in the magazine *Frakcija* in 1997 (Miščević, 1997). The same issue brought also a translation of Aldo Milohnic's essay entitled 'Performative theatre' (Milohnić, 1997), in which the author tried to apply the Austinian paradigm to the study of performing arts, and offered his typology dividing 'literary' from 'performative' theatre, the latter largely covering various instances of performance art, i.e., the art of Austin's 'happy performatives,' since it is only there that all its conditions seem to be duly fulfilled, however, constantly bringing theatrical conventions into question. Although it first appears like the author aligns himself with Derrida and criticizes Austin's 'ostracizing gesture,' as he calls it, with respect to the art of theatre and its uses of language it finally emerges that the introduction of his typology only confirms the validity of Austin's suggestions, at least when it comes to 'literary' theatre which – despite Derrida's protestations – suddenly again rightly represents an instance of the 'etiolation of language.'

10. *Gender Trouble* was translated into Croatian in 2001, but became an object of scrutiny much earlier. Dates of translations of theoretical works are rarely a reliable factor in understanding theoretical 'influences' in a milieu that is trained to read, think, and write in foreign languages, from 'officially' to 'unofficially' imposed ones – Latin, Italian, German, Hungarian, Russian,

and today English, the knowledge of French being perhaps the least enforced and therefore the most distinctive.

11. We have extracted these characteristics from the works of two performance-theatre groups that were active in the 1970s and at the beginning of the 1980s, *Kugla glumište* (Kugla theatre) and *Coccolemocco*, as well as two authors who are still active, Branko Brezovec, who was also the founder of *Coccolemocco*, and Damir Bartol Indoš, once a member of *Kugla glumište*. We do not mean to suggest, of course, that the whole production of performance-theatre can be reduced to two groups and two authors.

References

Biti, V. 'Institucionalizacija semiotike u domaći akademski život,' in *Trag i razlika – čitanje suvremene hrvatske književne teorije*, ed. V. Biti, N. Ivić, and J. Užarević (Zagreb: Naklada MD / HUGHZ, 1995), pp. 107–22.

Biti, V. *Pojmovnik suvremene književne i kulturne teorije* (Zagreb: Matica hrvatska, 2002).

Blažević, M. 'Director vs. Actor or Matula vs. Brezovec,' *Frakcija*, 20–21 (2001) 125–43.

———. 'Taming the Vague: Gavella's Theoretical, New Actor-Spectator,' *Frakcija*, 28–29 (2003a) 78-86.

———. 'Theatre Doubts,' *Frakcija*, 28–29 (2003b) 120–33.

———. 'A few notes on Branko Gavella and his theory of acting,' *Performance Research*, 10:2 (2005) 70–4.

Bošković-Stulli, M. '*Usmena književnost*,' *Povijest hrvatske književnosti 1* (Zagreb: Liber-Mladost, 1978), pp. 7–353, 641–51.

Brezovec, B. 'Nikad nisam sebe smjestao izvan centra – samo sam centar pomaknuo malo sa strane' (intervju), Quorum, XV:2 (1999) 138–63.

Čale, M. *Volja za riječ, eseji o djelu Ranka Marinkovića* (Zagreb: Zavod za znanost o književnosti Filozofskog fakulteta Sveučilišta u Zagrebu, 2001).

Čale Feldman, L. 'Image of the Leader: Being a President, Displaying a Cultural Performance,' *Collegium Antropologicum*, 1 (1995) 41–52.

———. *Euridikini osvrti* (Zagreb: Centar za ženske studije i Naklada MD, 2001a).

———. '"Tis a pity she's a whore": An actress and her doubles between postcommunism and posthumanism,' *Frakcija*, 20–21 (2001b) 94–111.

———. 'Cultural Memory Petrified: Theatre in the Service of Propaganda,' unpublished paper delivered at the IFTR congress 'Theatre and Cultural Memory – The Event between Past and Future' (Amsterdam 2002).

———. *Femina ludens* (Zagreb: Disput, 2005).

Čale Feldman, L., I. Prica and R. Senjković, *Fear, Death and Resistance – An Ethnography of War, Croatia 1991–1992* (Zagreb: Institute of Ethnology and Folklore Research/ Matrix Croatic, 1993).

Cvejić, B. and A. Milohnić. 'DI-SO/LO-NANT DU-O,' *Frakcija*, 26–27 (2002/2003) 90–7.

Cvjetičanin, B. and V. Katunarić. *Cultural Policy in Croatia*. National Report (Strasbourg: Council for Cultural Co-operation, 1999).

Gavella, B. *Hrvatsko glumište: analiza nastajanja njegovog stila* (Zagreb: Zora, 1953).

———. 'Socijalna atmosfera Hrvatskog narodnog kazališta i njegovi odnosi prema svom kazališnom susjedstvu,' *Rad JAZU*, 353 (1968) 5–54.

———. *Književnost i kazalište* (*Literature and Theatre*) (Zagreb: Matica hrvatska, 1970).

———. 'A Barely Noticeable Stitch,' trans. T. Brlek, *Frakcija*, 20–21 (2001) 5–6.

Handelman, D. *Models and Mirrors – Towards an Anthropology of Public Events* (Cambridge: Cambridge University Press, 1990).

Heathfield, A. 'Alive,' in *Live: Art and Performance*, ed. A. Heathfield (London and New York: Routledge, 2004), pp. 6–13.

Jukić, T. 'A Lasting Performance: Jane Austin,' *Links and Letters*, 6 (1999) 23–34.

Kirby, M. 'Happenings: An Introduction,' in *Happenings and Other Acts*, ed. M. R. Sanford (London and New York: Routledge, 1995a), pp. 1–28.

———. 'The New Theatre,' in *Happenings and Other Acts*, ed. M. R. Stanford (London and New York: Routledge, 1995b), pp. 29–47.

Lehmann, H. T. *Postdramatic Theatre*, trans. K. Juers-Munby (London and New York: Routledge, 2006).

Lozica, I. *Izvan teatra: teatrabilni oblici folklora u Hrvatskoj* (Zagreb: Teatrologijska biblioteka, 1990).

Marjanić, S. 'Acting Transformations, Maintaining Artlessness and the Politics of Metaphysics,' *Frakcija*, 20–21 (2001) 15–26.

Milohnić, A. 'Performativno kazalište,' *Frakcija*, 5 (1997) 49–54.

Milovac, T. *The Misfits: Conceptualist Strategies in Croatian Contemporary Art* (Zagreb: MSU, 2002).

Miščević, N. 'Što su performativi?,' *Frakcija*, 5 (1997) 42–5.

Phelan, P. *Unmarked: The Politics of Performance* (London and New York: Routledge: 1993).

Pristaš, G.S. 'Visiting Art,' *Frakcija*, 16 (2000), 30–9.

Sajko, I. 'Concrete Oedipus,' *Frakcija*, 22–23 (2001/2002) 126–37.

Vanden Heuvel, M. *Performing Drama / Dramatizing Performance, Alternative Theater and the Dramatic Text* (Ann Arbor: University of Michigan Press, 1991).

Vujanović, A. 'Terrible fish: Impossible dialogue,' *Frakcija*, 22–23 (2001/2002), 104–20.

Zlatar, A. 'Čistilište kazališnih teorija,' *Zor–časopis za književnost i kulturu*, 2/3: 2 (1996) 221–4.

Part III
The Power of Performance Practice

11
Performing Postcoloniality in the Moroccan Scene: Emerging Sites of Hybridity

Khalid Amine

The Maghreb is caught in an ambiguous compromise – at least linguistic – between Semitism and Latinity.

(Jacques Berque, 1983: 4)

Performance studies in Morocco has experienced a recent momentum through the establishment of the *International Centre for Performance Studies*, based in Tangier. The center is devoted to multidisciplinary research in performance studies at the national (Moroccan), Arab, and international level; dialogue and collaboration between artists and academics; the publishing of multilingual publications on theatre and performance studies; the organization of festivals, conferences, and workshops on theatre and performance studies; and the organization of the annual international forum, *Performing Tangier*. The center focuses on a number of research fields, including Theatre, Diaspora, and Emigration; Performing Postcoloniality; Cultural Diversity and the Making of Spectacle; and Performance Studies between East and West.[1]

One of the pressing problems that inform performance studies in Morocco remains the question: Did the Moroccans have a theatrical tradition of their own before the Franco-Hispanic colonialization? And what is the relation between this tradition and 'indigenous' performance forms? In the context of postcolonialism in Morocco, theatre is used by both the state and the political opposition as a means of empowerment or indoctrination. During the French-Hispanic Protectorates (1912–56), Moroccans had employed theatre as a weapon of resistance. Initially the colonial administration responded by banning theatrical activity, but decided in 1950 to render it more docile by producing massive popular entertainments that would smooth conflict. Professional experts were called from France, and their influence was felt even after independence in 1956. Until 1975, a period covering several postcolonial administrations, a collaboration was effected between the state and the stage in performing spectacles of power through adaptations from the Western repertoire. However, postcolonialism also brought with it emerging sites of resistance against such neo-colonial enterprises and the post-independence governments that sustained them. The 1970s and 1980s in Morocco are often referred to as the 'years of 'lead'; the country turned into a police state, reinforcing restrictions not only upon political opposition but also dissident artistic expression. Theatre played its part in the struggle. Professionals such as Tayeb Sadikki and Ahmed Tayeb Laalej have established a popular yet moderate theatrical tradition that deploys the techniques of the French Comedy as well as local performance behavior. Amateur theatre went into another direction, creating performances that were mostly progressive and revolutionary due to its adoption of pan-Arabism as a central ideology, and the Palestinian cause as a daily concern.

As a result, the debate about Morocco's theatrical tradition has oscillated between two positions. According to the first, Western theatrical models as represented in the proscenium tradition should be repudiated and replaced by a return to the 'indigenous' performance traditions (another way of returning to pre-colonial Morocco). This has led some to the worship of ancestors, and eventually to a (useless) quest for purity that can be called 'Arabocentrism.' Such essentialist enterprises rest upon a new myth of origin in the name of 'authentic' Arabic/Moroccan performance (an offshoot of Pan-Arabism in the realm of politics). The rejection of Western values through an uncompromising quest for authentic tradition has been reinforced by the deterioration of the economic situation in Morocco since independence in 1956. However, these so-called 'indigenous' performing traditions are in fact diasporic cultural constructs that constantly change and are transformed according to the inner dynamics of the fluid and yet adaptive folk traditions. Thus decolonizing Moroccan theatre from Western influence does not mean a recuperation of a pure and original performance tradition that pre-existed the colonial encounter. Such tendency falls into an inevitable process of essentializing and self-Orientalizing. As Chambers asks, 'Does there even exist the possibility of returning to an "authentic" state, or are we not all somehow caught up in an interactive and never-to-be-completed networking where both subaltern formations and institutional powers are subjected to interruption, transgression, fragmentation and transformation?' (Chambers, 1994: 74) There is no way back therefore to an 'authentic' state.

LIVERPOOL JOHN MOORES UNIVERSITY
LEARNING SERVICES

The second position sees the European proscenium tradition as a unique model that should be imitated and reproduced. However, Western theatrical models are more than dramatic/theatrical spaces, for they are cultural and discursive ones as well. Borrowing Western models without critiquing their exclusivist tropes amounts to a new kind of reification. In brief, this second position, which is held by many Westernized Moroccan critics and practitioners, falls into another kind of essentialism, which sees European theatre as a unique model that should be disseminated all over the world, even at the expense of other peoples' theatrical traditions and performative agencies. The Europeanization of Arabic performance (*Ta-awrub al-furja al-arabia*) exemplifies the complicity of colonized subjects within this.

I will attempt to highlight a third position between the two essentialist traps. My argument is that contemporary Moroccan theatre finds itself construed within a liminal space, on the borderlines between different tropes. It cannot exist otherwise, for it juxtaposes heterogeneous entities to emerge as a hybrid field that is spaced between East and West. It is a fusion of Western theatrical traditions and local Arabic performance practices. The hybrid nature of Moroccan theatre is, firstly, manifested in the transposition of the *halqa* (a public gathering in the form of a circle) from the public space of *jemaa-elfna* (a square in Marrakech) to modern theatre buildings like the National Theatre Mohamed the Fifth,[2] which reproduce the Eurocentric proscenium paradigm;[3] and secondly, it is evidenced in the application of a high sense of self-reflexivity.

Hybridity is not simply a fusion of two pure moments, but the persistent emergence of liminal third spaces that transform, renew, and recreate different kinds of writing out of previous models. As Bhabha suggests, '[I]f the act of cultural translation (both as representation and as reproduction) denies the essentialism of a prior given originary culture, then we see that all forms of cultures are continually in a process of hybridity, but for me the importance of hybridity is not to be able to trace two original moments from which the third emerges, rather hybridity to me is the "third space" which enables other positions to emerge' (Bhabha, 1990: 211). Even though Bhabha's approach is mostly textualist and theory-bound and lacks a practical counter-narrative, his concepts yield practical agency when applied to concrete focalized scrutiny. Moroccan theatre is informed by an intentional aesthetic hybridity as it tends to juxtapose different heterogeneous elements that belong to opposing performance traditions. The effects of this hybridity are manifested in its ironic double consciousness.

Al-halqa: street performance as social drama

Al-halqa is the most overtly theatrical among the public spaces in Morocco. It is a public gathering in the form of a circle around a person or a number of persons (*hlayqi/hlayqia*). As a site-specific popular entertainment, *al-halqa* hovers between high culture and low mass culture, sacred and profane literacy and orality. Its repertoire combines fantastic, mythical, and historical narratives from *Thousand and One Nights* and *Sirat Bani Hilal*, stories from the holy *Quran* and *Sunna*, and witty narrative and performative forms. The medium of the *halqa* varies from storytelling to acrobatic acting and dancing. *Al-halqa* still remains the most significant performance behavior not only in Morocco but throughout all the Arab World, as it is deeply rooted in Arabo-Islamic heritage. In all its diversity, *al-halqa* serves as a source of artistic delight and entertainment, and as a means of spatializing cultural identity. It is also a medium of information and circulation of social energy, a social drama.

Al-halqa contributes to the representation of historical consciousness and cultural identity through its formulaic artistic expression. Seen from a Bakhtinian perspective, these stories and narrative utterances are strongly affiliated with individual and communal lives and have been repeated often enough to become narrative performances that bear clear traces of an individual's sense of identity. Given *al-halqa*'s magical capacity to implicate passers-by through the energy that radiates all over its surrounding space, it negotiates the differing relationships among its participants, and in the process it reformulates cultural values and self-knowledge as it engages its audience in a constant game of role-playing. The performance imbues a human action with a heightened potential to shape, reflect, and mirror cultural identity. This aesthetically marked space is a constantly rehearsed oral text that is (re)written time and again under erasure through artistic expression, ranging from narrative folktales and storytelling to ritualistic dancing, theatrical pantomime, and improvisation. *Al-halqa* genuinely encompasses all these genres and representational practices in a single performance text, a text that is dialogical through and through since it is constructed as patterns of infinitely self-erasing traces. *Al-halqa*'s textual practice, however, transcends the boundaries of the written word as a scriptocentric closure, for it is a dynamic network of interrelated codes that are not necessarily linguistic.

Al-halqa has a managed environment that is strictly opposed to the Eurocentric closed theatrical institution. Its audience is called upon 'to drift' spontaneously into an arc surrounding the performance from all sides. The space required by the *hlayqi* (the maker of spectacle) is not a specific space, and the timing of the performance is any time. No fourth wall with hypnotic fields is erected between stage and auditorium, for such binary opposition does not exist in *al-halqa*. All the marketplace or Medina gates can be transformed into a stage, and the entire circle functions as a playing area, as open as its repertoire of narratives and dances. The morphology of the Arabo-Islamic city as manifested in most Moroccan ancient *medinas* (Fes, Marrakech, Taroudant, Tangier, Sale, etc.) shows that the circle is an essential paradigm in the refashioning of the city as well as the social imaginary of its inhabitants.[4] The *medina* is a circle surrounded by many gates wherein the Mosque is situated at the center as a spiritual icon, as well as a commanding and surveying cultural apparatus. Around the mosque there are other micro-circles that are organized hierarchically, from the most privileged artifacts, bazaars, and houses located near the center, to the poorer areas of the outer circles that face the gates. *Al-halqa* performance as a free and licensed expressive behavior is situated most often at the *medina* gates and market places (far from the sacred center and its sacred didactic *halqas*). It is thus a tolerated form of voicing the boundaries between sacred and profane. Though *al-halqa* is situated in the periphery of the circle, it functions as an entertaining social commentary – that amounts to parody sometimes – on what is going on inside the circle. It is a carnivalesque mirror of topsy-turvidom, producing, as a matter of fact, an oscillation between the inside out and the outside in of the whole circular medina.

Most of the micro-circles inside the medina are deployed as performative spaces, playing grounds, and sites of group experience and outdoor education for children and adults. While walking back and forth from the traditional Quranic School called *Lamssid* and their homes, the sites of two disciplinary apparatuses, children get familiarized from their early formative years with the environment of *al-halqa* as a liminal space which represents an ideal platform for spontaneous performance behavior, due to its licensed liberating spirit. The frequent visits of itinerant professional performers to the narrow streets of the ancient *medina* of Fes illustrate the potential efficacy of collective outdoor rehearsed activities. Mohamed al-Kaghat highlights one performer's omnipresence all over the city of Fes: 'The most famous visitor to the medina and perhaps the last one remaining from the entertainers *of Rass-Lklia* square is a man named Harrba [...]. He used to hang around all over the medina tuning his performance in accordance with his audiences and their number [...]. He is Fes' Tespus, a traveling actor who brings performance to people' (al-Kaghat, 2002: 30). As a professional *hlayqi*, Harrba turns the whole city of Fes into a site-specific performance.

In summary, *al-halqa* performing traditions are social dramas that encode a cultural behavior and 'enable people to understand themselves and their worlds, through the medium of their culture' (Hornby, 1986: 52). Through a performance studies approach we may understand these performance behaviors as self-reflexive representational acts that mimic a given presence. Schechner defines a performance behavior as a 'known and/or practiced behavior – or "twice-behaved behavior," "restored behavior" – either rehearsed, previously known, learned by osmosis since early childhood, revealed during the performance by masters, guides, gurus, or elders, or generated by rules that govern the outcome, as in improvisatory theater or sports' (Schechner, 1985: 118). Schechner thus enlarges the field of theatre studies by emphasizing the potential efficacy of performance activities and their impact on theatre practice: 'The ever-increasing use of outdoor public space for rehearsed activities – ranging from demonstrations to street entertainers – is having an impact on indoor theater' (Schechner, 1994). The shift during the last decades from theatre studies to performance studies thus presents a significant change from a rigid discipline committed to the textual world of drama and its manifestation onstage, into a hybrid field wherein expansive currents from various disciplines meet and which allows us to re-evaluate particular performing traditions.

Moroccan performing traditions as manifested in *al-halqa* are narratives loaded with conflict, besides being theatrical through and through; they are enacted repeatedly time and again, for they provide social commentary on contemporary Moroccan society in a self-reflexive way. These performing traditions used to occupy the space known today as theatre; yet a theatre that is free from the confines of logocentrism as manifested in the proscenium tradition that separates stage from auditorium and theatrical representation from its social context. Victor Turner observes that 'originally theater was concerned, among other things, with resolving crises affecting everyone and assigning meaning to the apparently arbitrary and often cruel-seeming sequence of events following personal or social conflicts' (Turner, 1982: 115). These performances have been profoundly reflexive social dramas, which were relegated to a defensive position through the forces of modernization.

The hybrid trope of the postcolonial turn

After a period of appropriating Western theatrical models (from the 1920s until the early 1960s),[5] Moroccan postcolonial dramatists established a dialogue with *al-halqa* and eventually transposed it to the space of dramatic writing. Tayeb Saddiki pioneered this new dynamics with a play entitled *Diwan sidi Abder-rahman al-majdub* (1965–66) (*The Collection of Master Abder-rahman al-majdub* – abbreviated as *al-majdub* hereafter). In the course of his career as Morocco's most established playwright and director, Saddiki has not only dominated Moroccan theatre, he has in fact established it as a tradition. *Al-majdub* represents the emerging festive theatrical enterprise in postcolonial Morocco; it is placed on the borderline between Western theatre and Moroccan performance traditions. For the first time in the brief history of Moroccan theatre, Saddiki transposed al-*halqa* as an aesthetic, cultural, and geographical space into a theatre building (transplanted into Morocco as a subsidiary colonial institution).

Al-majdub is a play conceived in an open public place. It refers us to its hybridized formation through its persistent self-reflexivity. The play's structure is circular rather than linear. It is situated in *jemaa-elfna* as an open site of orature and a space of hybridity itself. The first scenes of *al-majdub* attract our attention to the making of *al-halqa* and its circular architecture. Actors transcribe the circular form of *al-halqa* through a series of comic acrobatic games and mimetic body language. They play audiences to each other as the narrator (the storyteller) gives space to his little *halqa*. The *halqa* of *al-majdub* represents the Moroccan popular poet like a Shakespearean fool, giving voice to wisdom in a corrupt social order. The effects of such an absurd situation are comic yet redemptive, leading to a collective catharsis. The dramatic text and theatrical production of *al-majdub* are still considered as masterpieces and landmarks of Moroccan theatre, for they opened a new trajectory of possibilities and experimentation.

Maqamat Badia Ezzamane El-Hamadani (1997) is another turning point in Moroccan theatre, as it restores the performative and theatrical qualities of *maqamat*[6] and its narrativity back to the Moroccan and Arab stage. In the prologue, Saddiki admits that his present *bsat*[7] is no other than a restoration of the deeply rooted *halqa* tradition:

> In my endless quest for an original Arab and particularly Moroccan theatrical form, I've fond in Badiaa Ezzamane's *maqama*, written a thousand years ago, dramatic structures, intrigues [...].
>
> (Saddiki, 1998: 1)

The play takes place in an open public square. The performance lacks an organic thematic unity, for although all stories used are derived from the *maqamas*, they are fragmented into little *furjas* or *halqas* that have only one common aspect – the master narrator. However, Saddiki's fragmentation strategy is rooted in the dynamics of the *halqa* and its fluidity.[8]

Ahmed Tayeb Laalej has also made use of *al-halqa*'s verbal efficacy in many of his theatrical performances. His use of the rhymed language of *al-malhoun* and popular *nawadir*[9] is highly influenced by *al-halqa*'s dynamics of narrativity. His theatre has been established since the 1950s as a popular theatre that is open to French comedy, comedia dell'arte, and, most importantly, to the rich repertoire of *al-halqa*. In *Juha wa cha-jarat a-tufah* (*Juha and the Apple Tree*) (1998), for example, Laalej uses *al-halqa*'s techniques of telling and showing, and a story that stems from deeply rooted orality. The techniques deployed by Laalej mix up storytelling, which is represented by the *rawi*'s (reciter's) narration and comments on the action taking place onstage, and a dramatic line that unveils the events to an onstage audience as well as those in a possible outer ring. Themes of corruption and social injustice reverberate sharply for the writer. *Juha wa chajarat a-tufah* is an ironic representation of the deeply rooted corruption of elections in Morocco: The egocentric and hypocritical Juha uses whatsoever means available in order to convince his potential voters, but once elected and exalted onto the top of the fruitful apple tree, he disregards them.

Since Saddiki's *Majdub*, al-*halqa* has been a vital source of restoring tradition. Zober Benbouchta's twin plays *Lalla J'mila* and *Annar L-hamra*, both subsidized by the Ministry of Culture in 2004 and 2006 respectively, exemplify the dynamism of old ways being hybridized with new artistic venues. *Lalla J'mila* (first performed by Ibn Khaldoun Theatre Company in 2004) explores the underground history of Tangier as a social project that is governed by paternalistic systems, yet represented with feminine qualities as the 'Bride of the North' and the 'Pearl of the Strait.' The performance is an act of memorizing as well as a scrupulous practice of excavating layers of little histories and fragmented first-person narratives to reveal the interpenetration of space, culture, and gender. It unlocks histories of Moroccan sexual politics within a situation marked by colonial hegemony on the one hand, and the deeply rooted local patriarchal mindset on the other. Highlighting constructed polarities such as centrality and marginality, high culture and mass culture, masculinity and femininity, the play braids together local and global issues.

Within the space of *al-halqa*'s storytelling, which 'protects the weapons of the weak against the reality of the established order,' as de Certeau has suggested about narrative (de Certeau, 1984: 23), the story of two sisters become a means of empowerment when other means are denied or beyond their reach. It foregrounds female voices that are usually not publicly exposed. Ben Bouchta's play was written for, and co-produced by, an active feminist network in response to the implementation of a new family code, called *mudawanat al-ussra* (the Family Code) in 2003, which gives more rights to women regarding marriage, education, and child custody. In its persistence upon problematizing the old-fashioned division between public and private space, the play also searches for a better correlation between space and women's bodily existence; and in so doing, it calls for a complete shake-up of the paternalistic policies of the family.[10]

For almost half a century, *al-halqa* has thus been a vital source of energy in Moroccan theatre. Its retrieval and transposition to the stage exemplifies the potential of liminality both as intentional artistic hybridity and postcolonial agency. However, if repetition constitutes one of the fundamental aspects of the *halqa* as a 'restored behavior,' it has also affected its theatrical use. Most Moroccan plays inspired by *al-halqa* have turned it into a museum piece. The enormous potentialities of *al-halqa* remain unexplored. There are no records of any interdisciplinary experimentation with al-*halqa*'s repertoire of pre-expressivity, or with its acting techniques, which would exalt the actor's energy and presence. There are no records either of any effective use of *al-halqa*'s political praxis following Boal's model of the *Theatre of the Oppressed*. *Al-halqa* is still conceived within the age-old confines of actor versus spectator. Yet the circularity of the *halqa* has an ideal potential to transmit energy and make everyone into a participant.

In this context, the Algerian Abdelkader Alloula (1929–1994), who performed some of his plays in Morocco, remains the most significant Maghrebi artist, besides Saddiki, to use *al-halqa*. Alloula directed all his creative energy in the last decade of his life into developing a theatrical methodology that was drawn from the Algerian *halqa*, called *al-quwal*. He experimented with *al-halqa*'s techniques in his last five works: *Legoual* (1980), *Al-Ajwad* (1985), *El-lithem* (1989), *At-tufahu* (1992), and *Arlukan khadimu as-sayyedayni* (1993). Alloula expressed his concern with *al-halqa* at the tenth conference of *the International Association for Theatre Critics in Berlin* 1987, after he dismissed the theatrical arrangement enclosed within the Italian box: 'Aristotelian theatre has proved its limitation for peasants whose typical cultural behavior towards the theatrical exposition involves sitting on the floor and forming a circle around the theatrical arrangement [...]. The entire theatrical exposition must have been reviewed altogether' (Alloula, cited in Mnowar, 2005: 7),

This statement exposes the motives that led Alloula to focus on *al-halqa* during his mature period between 1980 and 1994. It is not so much an essentialist appeal seeking exile in the past and returning to an original Arab performativity, as an experimentation with performers and potential audiences. Through the estrangement of theatrical semiosis and its displacement from conventional dramatic construction – with the help of a circular setting and Brechtian acting techniques – Alloula's aim is not only to renew our perception of what is presented on the *halqa*'s stage, but also to reconstruct critically what has been already dismantled. Regarding this, Alloula's theatrical techniques are highly politicized. Alloula's theatre is rooted in situations marked by tension and extracted from critical historical moments. It highlights the contradictions existing between the individual and the collective in postcolonial Algeria.

Performing liminality

The current critical emphasis on self-referentiality brings to the fore an important component in contemporary theatre practice, namely its privileging of representation. The celebration of theatricality and performativity has become dominant not only in the Western tradition, but in the postcolonial Arab theatrical scene as well. This phenomenon has become a generalized feature of the so-called postmodern era of writing – as well as the postcolonial one – since our Global Village now is not only a space of 'mobile objects' but also one of 'reflexive subjects'; not only one of global hyper-commodification but of opportunities for enhanced social and intellectual cooperation. The ability to reflect upon social conditions of existence is linked to the process of globalization and to the accelerated de-traditionalization of societies – including developing ones like Morocco (see Beck et al., 1994). Hence, reflexivity becomes a characteristic feature of our age, which is embedded in our cultural processes and thus also in our theatre.[11]

Moroccan theatre tends to privilege this kind of self-reflexive performance since it is a theatre that is constructed within a liminal space. The effects of this hybridity are manifested in its double consciousness (a consciousness that is informed by the Western tradition, particularly French Comedy), as well as in its location between Self and Other, East and West, the local and the global, tradition and modernity, orality and literacy. These negotiations are also informed by Moroccan identity, an identity that is very much hyphenated insofar as it is Arabo-Berber, yet deeply rooted in Africa and open towards Mediterranean culture. Theatrical practice in Morocco thus emerges as a continuum of intersections and negotiations; the result of these is a complex palimpsest that highlights the powers of the hybrid and the impure rather than a logocentric quest for presence and purity. In other words, theatrical practice in Morocco is part of the dynamics of modernizing the country.

In order to rehabilitate a performance tradition, Moroccan theatre becomes more and more improvisational and self-reflexive, for such retrieval is still negotiated within the paradoxical parameters of appropriating and misappropriating the Western tradition. The self-reflexive theatre in Morocco deploys a twofold strategy of retrieval. The first one, which I have outlined above, focuses on retrieving a pre-theatrical performance tradition that is situated in rituals, ceremonies, and masquerades. The second strategy focuses on establishing a dialogue with the current theatrical scene elsewhere in the world. Such dialogue amounts to an auto-reflection whereby the mirror of theatrical representation no longer reflects an outside presence, but rather reflects a fragmented and over-hyphenated body that is as much contested as Moroccan identity is today. Comicality happens to be a common ground that brings the two tendencies together, yet with differences at the level of agency and reception, for the first tradition tends to absorb conflict through a compromising and redemptive comicality designed to cloud the public's inherent anger. The second, more fragmentary and self-reflexive tradition highlights conflict through a subversive comicality that uses comic effects as a means of dismantling the complacent passivity of the audience.

The comic is indeed one of the main means by which to cope with the Moroccan postcolonial situation in all its socio-economic and political dimensions. The redemptive tendency is manifestly present in popular theatre that uses conflictual situations in order to laugh at them (the productions of the National Theatre company led by Morocco's most popular comic actor, Mohamed El Jem, exemplify this approach). The subversive tendency highlights conflict in order to force the spectator not only to laugh at that which is unhappy (the mirthless laugh in Samuel Beckett's terms) but also to prompt him to act (a comicality with a Brechtian *Lehrstück* agency): Nabil Lahlou's *Ophelia N'est pas Morte (Ophelia is not Dead)* (1968), 'Abd al-Karim Birshid's *Utayl Wal-Kyal wal-Barud (Othello Horses and Gun Powder)* (1975), Youssef Fadel's *Ayam Laaz (Happy Days)* (1994) and *Yak Ghir Ana (Just Me)* (2003) are a few examples out of many.

Improvised theatrical projects have chosen to subvert temporarily the theatrical apparatus in Morocco, taking it to absurd and uncompromising extremes when it comes to the politics of reception. Mohamed al-Kaghat remains the best representative of the subversive Moroccan *murtajala* (improvised play). Al-Kaghat is an academic, playwright, director, and actor, who is well acquainted with the Western impromptu improvisational tradition and its ironic representation of the theatre, a tradition that goes back to Molière's *L'Impromptu de Versailles* (1663). Al-Kaghat realized the ability of such a genre to communicate through self-reflexive comicality and performativity. In his prologue to *Murtajalat Fes* (*The Improvised of Fes*) (1991), he legitimatizes both his practice and our need for it 'because our theatre suffers from all kinds of problems, I have adapted the *Impromptu* in order to expose them to the audience after I realized that discussing problems is not as effective as performing them on stage [...] Through Irony and Comicality and the exaggeration of comic situations I desire to create a dark comedy [...]' (al-Kaghat, 1991: 7). Thus, the improvised play becomes a legitimate theatre practice that is based on an unfinished script, to be filled in the process of the performance event through the actors' improvisation.

The *murtajala* is comic through and through, due to its hilarious witty dialogues, comic situations, and dramatis personae, and its sharp critique of theatre practice within its social milieu. Still, it is considered a dark comedy as it foregrounds the old Moroccan saying that 'more sadness makes you laugh.' In *Murtajalat Fes*, for example, through an ironic representation of a corrupt and ruthless judge, al-Kaghat reminds us of one of the most painful moments in Moroccan theatre's brief history, namely the Fakih Ahmed Ben Saddik's *fatwa* against the practice of theatre and acting at large:[12] 'Ah... Ah... You don't know that acting is forbidden by divine law? [...] You don't know that the imitation of non-believers is forbidden?' (al-Kaghat, 1991: 83–4) The statement sums up a whole mindset that still regards the mimetic practice of theatre as an evil that should be eradicated from Arabo-Islamic culture. Al-Kaghat's ironic reflection on the subject illustrates the problems that hinder artistic expression in Morocco. Yet, in *murtajalat Chmisa Lalla*, the public's incessant pursuit of trivialities changes all of a sudden into a state of deep sorrow. Lalla Chmisa, daughter of the sultan, can no longer laugh or enjoy the beauty of life. So the sultan asks all actors and entertainers of the country to restore her smile and discover the causes behind her sadness. The play critiques the reification of theatre practice under the auspices of the government (especially the amateur theatre of the 1970s),[13] and reveals the impotence of the jury in the National Festival of Amateur Theatre. *Lajnat al-hukàm al-hukamaa al-muhanàkin* (the Committee of the Wise and Fat Jury) is supposed to be the savior of Lalla Chmisa. But instead, they deepen her sadness. The representation of the committee's debates reveals their theatrical illiteracy and inability to appreciate substantial art.

The same self-reflexiveness is manifested in *Imta nbdaw imta? (When are we going to start, when?)* (1997).[14] The performance is based on blurring the boundaries between drama and reality, acting and improvising, the actor and the character. The play starts with a nightmare experienced by an actress as part of her rehearsals, and suddenly the acting is interrupted by the director who emerges from the auditorium. The whole play then oscillates between the miserable life of the actors and director and that of the characters in their drama. Shakespeare figures in one of the dramas that are being rehearsed – an actress can no longer play the part of Ophelia with the same energy as in the old days, reflecting the miserable conditions of theatre making in Morocco. Meanwhile, the theatrical company is unable to find a venue, for they are ordered to move out time and again. The play becomes a reflection on the crisis of Moroccan theatre, a crisis that is caused by censorship and a lack of theatre buildings. However, the play also critiques the state of hesitation that characterized the mid-1990s in Morocco, just before the government of *tanawub* led by Abderrahman Youssefi was installed in 1998. The period was dominated by constitutional reforms, the release of many political prisoners, and above all dialogue between the monarchy and the leading opposition, yet the dynamics of reform and initiation of a democratic process in the country was determined by a sense of procrastination. The play displays the Moroccan political situation as still imprisoned within a vicious circle that does not transcend the waiting room. It is a call for action, summed up by the question in its title, 'When are we going to start?'

The recent *Murtajalat A'dar Al-Baydae (The Improvised of Casablanca)* (2003) by Masrah Adifa al-ukhra (The Other Bank Company) stages the predicament of theatre practice in Morocco during the present period after 1998 (and the installment of Mohamed al-Achaari as Minister of Culture), which is often called the 'period of change.' Through a deployment of black humor, the play dismantles the hegemonic structures that control theatre practice in a country that still regards artistic expression as a luxury rather than as a vital part of the construction of cultural identity. The play's comicality invokes a bitter laugh at the absurd situation in which these young actors trained at the *High Institute of Theatre* (ISADAC) in the capital city of Morocco find themselves. Through an ironic representation of the National Theatre's previous director and his naïve understanding of the needs and demands of acting professionals, the play sharply critiques government policies on theatre and calls for an urgent change. Again, the improvised form offers a means to make a statement through creating absurd extremes. Unlike *Imta nbdaw*'s optimistic call for action, *Murtajalat A'dar Al-Baydae* (conceived within the period of change) remains skeptical about the possibility of change. The political importance of the play lies in its depiction of present-day Morocco as essentially the same. The more this situation changes the more it remains the same.

In summary, Moroccan theatre today is constructed within a liminal space, on the borderlines between different tropes. It cannot exist otherwise, for it juxtaposes different heterogeneous entities only to emerge as a hybrid performance that is placed between East and West. It is a fusion of Western theatrical traditions and the local Arabo-Tamazeght performance traditions. The hybrid nature of such a theatre is manifested in the transposition of the *halqa* from open squares to modern theatre buildings, and in its emphasis on self-referentiality. Thus, the postcolonial condition of Moroccan theatre today is characterized by hybridity as a dominant feature. Hybridity is not simply a fusion of two pure moments, but the persistent emergence of liminal third spaces that transform, renew, and recreate different kinds of writing out of previous models. This constitutes our performance difference.

Notes

1. See *International Centre for Performance Studies website*: http://www.icpsmorocco.org.
2. Upon the independence of Morocco in 1956, the National Theatre Mohamed the Fifth was created in Rabat under government auspices, along with the Moroccan Theatre Research Center.
3. For a more extensive discussion of the translation of *al-halqa* from open space to theatre building, see Amine, 2001.
4. *Al-halqa*'s circular form is also manifested in the nomadic life of Moroccan peasants living in the *duwar* (which literally means the circle). In medieval Moroccan society, a *duwar* was the circle of tents of nomads whose cattle was kept inside the circle in order to be well supervised. The tradition of *tawaf* around the holly Kaaba during pilgrimage to Mecca also highlights the divinity of the circular form in Islam. In the Sufi conception of *al-hijab* (the veil), and particularly in Ibnu Arabi's symbolic reading of Arabic alphabetical graphic form, the circle stands midway between divine order and the human order. According to Ibnu Arabi, the letter *nun* (n) 'is half a circle. And the demarcating point that is located on top of the letter reveals the significance of the other nun, that is the divine nun (the sublime *nun* as opposed to the underground nun). Also, the nun underwrites the circularity of the universe' (cited in Belkacem, 2000: 49). The divine half of the nun remains unseen, and so is the divine order that is veiled for the general public. To unveil such order one needs to undergo a whole redemptive process of self-annihilation

and learning. In accordance with the Islamic experience of divinity and *al-hijab*, Arabo-Islamic arts have generally privileged the Arabesque tradition and its circularity and openness.

5. Morrocco became independent in 1956.

6. The *maqama* (or assembly) is a long narrative poem. The tradition of maqama started in the eleventh century when Badi' al-Zaman al-Hamadhani composed his first *maqamat*. Though it has dramatic characteristics, the *maqama* cannot be regarded as a complete play destined for the stage. The *rawi* (narrator) presents his narrative in the form of storytelling, yet adapts different roles to render his characters more flesh and blood. However, the poetic aspect of the *maqama* is much more dominant than its theatricality.

7. *L'bsat* is a performance tradition similar to conventional theatre. It is based on a managed stage and scenography, and most importantly, it has archetypal characters (see Amine, 2001: 65).

8. The most recent *bsat* performance in Morocco is a play entitled *Lbsaytiya* (the *bsat* people), performed in the National Festival of Theatre, July 2006 by one of Marrakech's most prominent theatre companies (Warchat Ibdae Drama). *Lbsaytiya* reproduces the same spirit of fragmented *furja*. The whole performance is inspired by the magical spell of *Jemaa El-fna*, as *lbsaytiya* agree to present their various *bsat* performances in Marrakesh's most famous square.

9. Plural of *nadira*, an extraordinary tale

10. For a detailed discussion, see Amine, 2007.

11. Richard Hornby suggests that: 'Metadrama can be defined as drama about drama; it occurs whenever the subject of a play turns out to be, in some sense, a drama itself. There are many ways in which this can occur [...] 1- The play within the play, 2- The ceremony within the play. 3- Role playing within the play. 4- Literary and real-life reference. 5- Self-reference' (Hornby, 1986: 31–2).

12. Ahmed Ben Saddiq (1889–1946) was a Moroccan scholar of Islamic Studies whose letter, entitled *iqamatu a-d-dalili 'ala hurmati a-t-tamtili* [Establishing Proof Against Acting], published in Qairo in 1941, dismissed all activities related to acting and theatrical representation. Among his arguments is that 'theatre leads women to prostitution, for there is no respectful actress since women are irrational beings by nature' (in Bahraoui, 1995, p. 7). Ben Saddiq's argument reveals prejudice not only against theatrical activity but also against women.

13. During the 'Years of Lead' in the 1970s (see above), freedom of expression in Morocco was controlled, and theatre was utilized by the political left to voice popular anger and its solidarity with the Palestinians. The amateur theatre festivals were also sites of politicizing theatre through a number of manifestos, through which the political sometimes dominated aesthetic concerns.

14. The play was originally written by Algeria playwright Mohamed ben Kentaf, and adapted and edited for the stage by Abdelouahed Ouzri and his team in Masrah al-yawm (Today's Theatre), a team that also comprises the distinguished Moroccan actor, Touria Jabrane, the playwright/actor and scholar Mohammed al-Kaghat, and the Moroccan singer Bashir Abdou. All references are from the televised production of the performance that took place in Tangier, 10 December 1997. Masrah al-yawm is a politically committed theatre company; its leading members are affiliated with the Union National des

Forces Populaires (UNFP), one of the leading socialist parties of the country. From 1998 to 2007, the then Minister of Culture, Mohamed Al-Achaari, belonged to that party. Touria Jabrane, the leading female actor of Masrah al-yawm, has been the Minister of Culture in Morocco since September 2007.

References

al-Kaghat, M. *Chmisa Lalla* (unpublished script, n.d.).

———. *al-murtajala al-jadida & murtajalt Fes (The New Improvised Play & The Improvised of Fes)* (Unfinished Theatrical Projects) (Casablanca: Sabou Publications, 1991).

———. *al-mumatil wa-alatuhu (The Actor and His Machine)* (Rabat: Ministry of Culture Publications, 2002).

Amine, K. 'Crossing Borders – Al-Halqa Performance in Morocco from the Open Space to the Theatre Building,' *The Drama Review (TDR)*, 45:2 (2001) 55–69.

———. 'Performing Gender on the Tremulos Moroccan Body: Zoubeir Ben Bouchta's Lalla J'mila,' *The Drama Review (TDR)*, 51:4 (2007) 167–73.

Bahraoui, H. 'al-islam wal-masrah,' *Alamat*, 4 (1995) 7–15.

Beck, U. et al. *Reflexive Modernization: Politics, Tradition and Aesthetics in the Modern Social Order* (London: Polity Press, 1994).

Belkacem, K. *al-kitabatu wa-ttasawufu inda ibnu arabiy* (Casablanca: Tubkal, 2000).

Bhabha, H. 'The Third Space, Interview with Homi Bhabha,' in *Identity, Community, Culture, Difference*, ed. J. Rutherford (London: Lawrence & Wishart, 1990), pp. 207–21.

Berque, J. *Arab Rebirth: Pain and Ecstacy*, trans. Q. Hoare (London: Al Saqi, 1983).

Chambers, I. *Migrancy, Culture, Identity* (London and New York: Routledge, 1994).

de Certeau, M. *The Practice of Everyday Life*, trans. S. Rendall (University of California Press, 1984).

Hornby, R. *Drama, Metadrama and Perception* (London and Toronto: Associated University Press, 1986).

Kaye, J. and Zoubir, A. *The Ambiguous Compromise: Language, Literature and National Identity in Algeria and Morocco* (London and New York: Routledge, 1990).

Laalej, A. T. *Juha wa Chajarat A-ttufah (Juha and the Apple Tree)* (Tangier: Chirae, 1998).

Mnowar, A. 'Tawdif aturat achabi fi al-masrah al-jazairi' ('The Use of Popular Heritage in Algerian Theatre'), *Annoormagazine*, 156 (December, 2005) 1–6.

Saddiki, T. *Sidi Abderrahman Al-majdub* (unpublished script, no date).

———. *Maqamat Badiaa Ezzamane El-Hamadani (An Entertaining b-sat)* (Kenitra: Boukili Publications, 1998).

Schechner, R. *Between Theater and Anthropology* (Philadelphia: University of Pennsylvania Press, 1985).

———. *Environmental Theatre* (New York: Applause, 1994).

Turner, V. *From Ritual to Theatre: The Human Seriousness of Play* (New York: Performing Arts Journal Publications, 1982).

12
Searching for the Contemporary in the Traditional: Contemporary Indonesian Dance in Southeast Asia

Sal Murgiyanto

The focus of this chapter is on the relationship between the 'contemporary' and the 'traditional' in the way dance is approached in Indonesia – and indeed, in a number of Southeast Asian contexts. This 'approach' is itself a form of research. Two facts are of particular salience here. First, in Indonesia, most choreographers are trained in traditional dance. There are some choreographers who are trained in Western forms such as ballet and modern dance, but they are in the minority. Second, it is worthwhile noting that there is no single Indonesian culture. The country consists of many ethnic cultures, and therein lies one of Indonesia's strengths – its possession of a rich and diverse variety of traditional artistic forms. Despite these two facts, Indonesian choreographers are interested in contemporary dance, but this orientation is distinct in that it does not forgo traditional values and the traditional dance forms.

The formation of contemporary dance, then, comes about as an *approach* to dance creation by Indonesian dance artists and choreographers, and not as the attempt to create a specific style, since the resulting form is open both to individual styles and various traditional cultures. Contemporary Indonesian dance practice and choreography, then, become forms of research – of practice as research, even if not in the formal sense of that term – in which there is a dialogue with respected traditions. This is an important cultural development of how a model of 'becoming modern' is formed, given a certain desire for the traditional.

This chapter, as such, explores the unique position of Indonesian dance and, indeed, dance artists in Southeast Asia, and their particular creative approaches in creating contemporary work through the way collaboration as a type of protocol enables dance artists to (re)develop their respected dance traditions; and how, further, through this means, modernity is interpreted and, as it were, put into practice. This creative approach, then, gives us the focus of this chapter, and the approach is not one that is simply locked into the experience of practice. It is not possible, though, to cite example of works from the 499 ethnicities that comprise the Indonesia nation. I will instead document instances of contemporary work that are taken from the three main Indonesian islands of Java, Bali, and Sumatra.

Collaboration as a trend in contemporary dance

Collaboration is a noticeable regional trend in dance creation. In July 2004, Sam Sathya, a Cambodian dance-teacher from the School of Dance, Royal University of Fine Arts, participated in a two-week choreography workshop organized by the Indonesian Dance Festival (IDF) VII of 2004 in Surabaya. At the end of the workshop, in collaboration with an Indonesian choreographer from Jakarta, she created a beautiful piece performed at the Emerging Choreographers Forum of the Festival. Two years later, she recalled her meaningful creative experience:

> When I first had the chance to explore contemporary dance, I was nervous. It seemed impossible for me to do it since the movements of the contemporary dancer I first saw were so inconsistent to my Cambodian classical dance training [sic]. However, my participation in creating contemporary dance with [Indonesian] artists [...] gave me greater confidence. What most impressed me was that these particular artists created contemporary dance based on their national traditional dances, which I thought was great. Though it is contemporary, it has national identity. The experience helped me to look out of the box and gain a sense of how dance forms can be developed.
>
> (Sam Sathya, 2006: 11)

Two important lessons are evident in Sathya's statement: first, the important role of tradition in practitioners' research in, and creation of, Indonesian contemporary dance; and second, the significant values of collaboration to open up one's mind and stimulate creativity. In 2006, two other Cambodian dancers were sent to participate in the IDF VIII Choreography Workshop.[1] Collaboration, as I have noted, is a significant means for 'thinking' and 'imagining' a contemporary dance, one still related to established 'national traditional dances' and 'classical dance' training. The national, the traditional, and the contemporary are brought together in a practical act of research.

On Tuesday, 18 July 2006, upon returning from Surabaya, East Java, to select five works of the ten-day choreography workshop to be presented at the IDF VIII's Emerging Choreographers Forum in Jakarta, I received a message from Indonesian choreographer Sardono W. Kusumo. He asked me to go straight from Jakarta airport to West Java, so as to observe his latest, 42-minute-long collaborative work entitled *Sunken Sea* performed at Selasar Sunaryo Arts Space (SSAS) in Bandung.

This performance was arranged for participants of the International Conference on Philosophy, 'Culture and Civilization,' organized by the Parahiyangan Catholic University in Bandung. SSAS – designed specifically and set on the hilly area in the outskirts of Bandung – is an independent art space that belongs to the famous sculptor-visual artist Sunaryo. For this performance, Sunaryo built a huge site-specific installation, *The Mountain of Wind*. This is a white-box structure (approximately $6 \times 15 \times 4$ meters) with a pair of palm trees on the top that slants down at about 30 degrees, and built diagonally on top of the SSAS's Amphitheatre. SSAS curator Agung Hujatnikajennong tells us of this collaborative effort:

[This] performance has been specially designed as part of an interdisciplinary collaboration between the two Indonesian great artists, both of whom have long addressed their works and projects toward the same personal concerns, particularly on the discourses of local tradition and wisdom as well as ecological problems resulted from an imbalanced modernization. This project marks a second collaboration between the two [Sardono and Sunaryo] and [is] a significant milestone of their artistic counterpart after more than three decades of respective career [*sic*].

The *Sunken Sea* dance performance and *The Mountain of Wind* site-specific installation are altogether constructed as an expression of concerns toward recent natural disasters happened in Indonesia and articulate our deep distress for the victims [...]. Within the context of the performance and installation by Sardono and Sunaryo, this is a moment to reflect back on [the] local wisdom that has taught us to keep aware and receptive in interpreting the erratic sign of nature.

(Hujatnikajennong, 2006: 2)

Sunaryo himself says, about his site-specific installation and collaboration, that:

This installation work is not made merely as a 'stage' for the sake of Sardono W. Kusumo's performance, but rather, to realize an idea constructed through my extensive dialogue with the dancer [...]. Undeniably, most of my works always depart from a search of the genuine meaning of nature's continual changing process – this time with a reference to the activity of Earth's Womb: the crack of the earth's plates and mountain's eruption resulting in the Tsunami swell, earthquake, landslides and other disasters for human being. To me all of these phenomena are caused by the age of the earth itself, which will ceaselessly grow along with its natural process and cannot be compared to the strength of creatures that live on it.

(Sunaryo, 2006: 6)

In a similar vein, after a long examination of different kinds of natural disasters occurring in the Indonesia's islands (Sumatra, Java, Bali, Flores, Kalimantan), and how local people wisely face and manage the disasters, choreographer Sardono concludes: 'It now seems important to consider a procession made up of myriads of cultural forms found throughout the Indonesian archipelago as an important expression of local wisdoms in reading the dynamic signs of nature' (Kusumo, 2006: 5)

Sardono Kusumo is a pioneer, the leading figure in Indonesian dance. His basic training is in classical Javanese dance, but he is well exposed to various traditional Indonesian dances from different islands (Bali, Kalimantan, Nias, and Papua) and also to modern Western dance. In contrast to Sam Sathya and the many Choreography Workshop participants at the IDF VII who are only starting to create contemporary dance, Sardono is a mature choreographer with more than 30 years of performance and choreographic career. As such, both his choreographic and general performative approaches and performance, which incorporate collaboration, are significantly different from emerging choreographers at the IDF. Performing in *Sunken Sea* with another Javanese male and two female dancers trained respectively in Sumatran (specifically Aceh-nese) traditional dance and classical ballet, Sardono and his dancers focused on the inner spirit, of being in total immersion and awareness of their dancing at the moment. In other words, for Sardono what is essential is not the form – the shape of the movement – but the *concept* and the *process* behind the movement. In *Sunken Sea*, he is concerned with the devastating natural disasters – tsunami and earthquake – that on 26 December 2006 destroyed some Indonesian regions and the importance of local wisdom in knowing how to deal with such disasters.

In Indonesia, collaborative work as a type of practice research that leads to the creation of new contemporary dance work has been in existence for more than a decade. In 1990, Balinese dancer-choreographer-scholar

I Wayan Dibia initiated collaboration with American dancer-musician-composer Keith Terry to create *Body Tjak*. This work tries to creatively bring together the rhythmic patterns of two performance forms to create new music patterns. The first form is Balinese *kecak*. *Kecak* can be defined as a multi-layered vocal music composition combining different rhythm of voice patterns *cak-cak-cak*. It was originally performed by male chorus to accompany the Balinese ritual of *Sang Hyang Dedari* (meaning 'a heavenly nymph descends to the earth'), to drive away evil spirits. In the 1930s, on the suggestion of German painter Walter Spies, I Wayan Limbak of Bedulu made a secular *kecak* composition recounting an episode in the Ramayana epic. *Kecak* is combined with simple hand and body movements, creating a magical atmosphere. These movements are drawn from American Body Music, a new music performance invented by American composer Keith Terry, in which rhythmical sound is produced through one's body by beating hands, arms, thighs, hips, chest, or by stomping one's feet to the floor, clapping, breathing, hissing, and shouting.

Regarding this collaboration, I Wayan Dibia wrote:

> From the beginning we both agree not to make *Body Tjak* a cut and paste work by bringing together finished-elements of either the *Body Music* or *Kecak*. By developing essential – more importantly rhythmical – elements of both art forms, *Body Tjak* must really be a new compositional form. By thoroughly mastering and understanding the rhythmical patterns, song's melodies, and movement elements from both forms, we collaborate our to the best of our ability to create movement and music patterns characteristic of *Body Tjak*.
>
> (Dibia, 2005: 379)[2]

Dibia believes that good collaboration must be done with mutual respect to the cultural and artistic values of the co-creators, as well as mutual understanding towards the artistic skills, knowledge, cultural attitudes, and beliefs of the participants. 24 performing artists – 12 from Bali and 12 from the United States – were involved in this 1990 collaboration. The artists were divided into two groups for the work, each composed of six women and six men. *Body Tjak* was reworked twice, once in 1999 as *Body Tjak, the Celebration*, and in 2002 as *Body Tjak Los Angeles*, with different numbers of participants and choreographic emphases. Dibia has noted that while the 1990 collaborative work could be called a theatrical form, the 1999 version was intended as a musical form to celebrate multiculturalism. As such, even though only 12 participants were involved in the latter collaboration, they were very multi-faceted, in ethnic-cultural terms. The six participants from Indonesia comprised three Balinese, two West Sumatrans, and one Sundanese artist, while the six American artists comprised Caucasians, African-Americans, and a performing artist from Cuba.

Such multicultural and, indeed, intercultural collaboration in artistic practice-as-research in the creation of contemporary art in dialogue with traditional cultural forms occurs not only in Indonesia, but also in other Southeast Asian countries such as Singapore, Malaysia, and Thailand. In 1999 in Jakarta, I saw Singaporean theatre director Ong Keng Sen's adaptation of Shakespeare's *King Lear*. This intercultural contemporary work, which was commissioned by the Japan Foundation Asia Center, was simply entitled *Lear*, and featured actors, dancers, and musicians from six Asian countries: Japan, China, Malaysia, Singapore, Thailand, and Indonesia.

In *Lear*, Indonesian choreographer Boi G. Sakti choreographed the movements of the characters called the Loyal Attendant, the Retainer, and the Warriors, through the adaptation of the West Sumatran traditional martial arts form, *pencak silat*. The accompanying music composed by Indonesian composer Rahayu Supanggah was informed by Javanese *gamelan*. Ong puts the various contributing cultures together in one performing space, but did not create a performative amalgamation which reduced cultural difference. Indeed, *Lear* performers, individually or in groups, speak or move in their respective dance language or dance style. Yuki Hata, the producer of *Lear* commented that 'Ong sought to reinvent the traditional art forms from a contemporary perspective' (Hata, 1999: 11). Ong himself has observed that such intercultural work 'negotiates roots, identity, and tradition' (Ong, 1999: 9) – such work requires 'practical' research into what tradition 'is' in the contemporary world.

In another Southeast Asian country, Malaysia, choreographer Marion D'Cruz has conducted many experiments with her fellow Malaysian artists – traditional as well as contemporary – connected with various artistic disciplines. Dance, music, visual, and media artists, supported by other non-artistic specialists, create contemporary works that opened up new physical, intellectual, spiritual, and sensory spaces (see D'Cruz, 2003: 83).

In Paris, I saw a collaborative work between Thai dancer-choreographer Pichet Klunchun and French conceptual choreographer Jerome Bel.[3] Five months later, on 3 June 2006, I again saw the collaborative work *Pichet Klunchun and Myself* in Taipei.[4] This piece deserves attention because it helps us understand what the key issues at hand are – collaboration as a trend, developing dance traditions, and the interpretation of modernity – and how they are engaged with in a common way by Southeast Asian artists. *Pichet Klunchun and Myself* is a performative and multicultural 'discussion-exhibition' between two artists, in which different traditions are engaged along a presumed East/West axis that is deconstructed even as it is also affirmed through each dancer's dance display of, and verbal explanation of, his art. In a sense, the 'research' each artist has put into the practice of his own dance tradition is put on display. The discussion-exhibition also represents what could happen in an intercultural interaction. The collaborative work deliberately 'displays' how even when two parties try their best to communicate, misunderstanding might still occur.[5]

Basically, the performance is divided into two parts. In the first part, Jerome Bel puts different questions to Pichet, and asks him to demonstrate and teach Bel different aspects of classical Thai dance. In the second half, it is the other way around: Pichet interrogates Bel and asks him to show examples of his work and explain the concepts behind it. At one point, Bel asks, 'What do you dance?' Pichet answered, 'I perform *khon*, a classical masked dance drama that combines Thai theatre, music, and dance, and used to belong to the Kings of Thailand.' He explains further that the story in *khon* is taken from the famous Indian epic, the *Ramayana*, and that King Rama I of Thailand (re)wrote the myth. King Rama II adopted *khon* as the official performance of the court which. This began around 220 years ago and continued until the reign of King Rama VI, who himself performed in the dance-drama as the *Ramayana*'s King Rama. Bel then asks Pichet to demonstrate various movements of different *khon* characters (human – both male and female, demon, and animal, specifically a monkey). Pichet then sings, shows how 'dialogue' between some characters takes place through movement, enacts a violent death scene, and teaches Bel some classical Thai dance movements.

If collaboration is one means through which contemporary dance in Indonesia is created, collaborative contemporary art – as we have seen – also occurs in many other Southeast Asian countries. Another way to create a contemporary work is through experimentation. The difference within the region, in this regard, though, is that in Singapore and the Philippines, where 'the growth of contemporary dance feels very connected to the progression of modern dance in America [...] and most, if not all, of the contemporary dancers come from a ballet background' (Lim, 2002: 4), in contrast, in Indonesia, Thailand, Malaysia, and Cambodia, most experimentations is geared toward developing, reinterpreting, or modernizing local traditions.

Searching for the contemporary in the traditional

I have thus far examined how the collaborative process helps enable a certain type of 'research' practice in the creation of contemporary dance. To discuss tradition and modernity (or the modernizing process in artistic creation), and the connection between the two here, though, it is necessary to discuss how we understand and define culture. Indonesia is a multicultural society. As such, following Bambang Sugiharto (2005), it is best to see culture not as a closed value system or a frozen pattern of behavior, but as a dynamic process, one that allows exchange and is part of complex historical interactions. The global cultural interaction bringing along with it 'modernity' and 'postmodernity' – hard to define as those phenomena are – has also wrought complex transformations in which new beliefs and value systems are no longer congruently related to the older ones. Old ideals are challenged by new ideas, practices, and lifestyles. In intercultural interaction, culture should be understood as a process of transaction, dialogue, and transformation, one in which various conceptual networks are rearranged, and in which previous forms of symbolic meanings are deconstructed and reconstructed, and we get new symbolic meanings and forms of thinking, seeing, and feeling, and also a changed self-awareness itself (see Sugiharto, 2005: 336–7).

In this respect, as Sugiharto further argues, tradition, then, can be defined as an inner struggle to give meaning to life experiences in a particular place and time, and as a systematic effort to understand the flow of events and the 'multi-' forms of world constellations. For most of us living today, 'tradition' has become an inner logic or pattern of perception, feeling, or imagining which is now hidden. The problem is that while in the past tradition took control of individual reflexivity, in our modern life it is the other way around: individual reflection controls and remakes tradition. Indeed, modernity has exposed us to other kinds of tradition in which ours is but only one. Modernity, on the one hand, is a continuous process of translating 'the otherness' into our language, horizon, biography, and into our consciousness and unconsciousness; on the other hand, it is a process of translating our horizon into the other horizons (see Sugiharto, 2005: 337).

The ones who have the potent capacity to open the inner logic of tradition hidden in the background of our everyday life are artists – especially performing artists. In their hands, the traditional past can be given a new soul, translated into a new context and horizon, and finally, transmitted dramatically and effectively to awaken consciousness, imagination, and feelings. Performing artists are also capable of bringing various daily struggles and phenomena into dialogue with traditional wisdom. In the hands of artists, a forgotten inner cultural vocabulary of the past can be revived and communicated in a fresh, new, surprising language of meaning that remain authentic and perceptive (see Sugiharto, 2005: 338). This is the context in which I write about the 'search' for the contemporary in the traditional.

In Indonesia, as earlier noted, contemporary dance production is a recent phenomenon. It began with the establishment of the Taman Ismail Marzuki Arts Center in Jakarta in 1968, when various Indonesian traditions were brought into dialogue with each other and with the diversity of Indonesian dance traditions, and also with the traditions of modern Western dance. Given the recentness of this cultural-historical development, how then can we understand what 'contemporary art' is, in the light of this chapter's concerns? Looking again at the dialogue in *Pichet Klunchun and Myself* may help us at least understand the different reasons for the various practices of contemporary dance in the West and Southeast Asia. Jerome Bel comments in the performance that:

> In the West, many people are interested in contemporary art. Contemporary art is an art that represents the reality of today and keeps in search of new forms. Contemporary artists try to produce arts that are 'new,' and art that they don't know. For this purpose some art producers give [grant] money to selected artists without knowing what kind of work they are going to see. That is why people come to see contemporary art. If they want to see what they already know, then they go to see ballet. That is why, I guess, 95 percent of production in contemporary art is very boring. But generally audiences still come to see contemporary art because they are curious. People go to see a contemporary performance and get bored. But again and again they come to see. Sometimes they see a good one [that is, a good production] and never forget [it] for the rest of their lives.

Contemporary art, then, can be said to represent the possibilities of 'the new,' For many Asians, given the historical legacy of European colonial domination, to 'modernize' effectively was to 'Westernize.' It is only natural that in some Southeast Asian countries, such as Singapore and the Philippines, 'the growth of contemporary dance feels very connected to the progression of modern dance in America' (Lim, 2002: 4). But for well-known Indian poet Rabindranath Tagore, 'True modernism is freedom of mind, not slavery of taste [...]. It is science, but not its wrong application to life' (Tagore, [1917] 2000: 355). Tagore discriminated between

Westernization and modernization, meaning that Asian modernization need not necessarily follow the patterns of modernization as it developed in the West. Not every Asian, of course, historically agreed with Tagore's position. The reception of his idea in early twentieth-century China was critical, yet Tagore never gave up on the possibilities of intercultural friendship (see ibid.: 355).[6] Yet Tagore also has strong contemporary supporters. Coming from Thai tradition and cultural background, Pichet Klunchun is one example:

> I want to invent my own dance movements that are still connected to Thai tradition. I went to the West and studied different dances, then went back home to Thailand to explore my own tradition. Many other Asians went to the West and have not gone back! I want to design my own life, to understand Thai culture. I am now 35 years old. Can't I design my own life and have my own dance?

Experimentation remains one means choreographers of contemporary dance may draw upon 'to move into areas they might otherwise not venture into' (D'Cruz, 2003: 77). This is the essence of contemporary art. For this very reason, many contemporary Asian artists have to break boundaries, existing rules, conventions of practice, and/or tradition to create a work that is 'distinctly exciting in form and content and profoundly meaningful for participant and viewer' (ibid.: 83). Creating the contemporary out of the traditional or looking into the future by exploring the possibilities in the rich sources of local traditions is a common cultural practice in Thailand, Malaysia, and Indonesia. One will find many examples of Tagore's version of cultural modernism not only in Thailand but also in India, Malaysia, and Indonesia.

Critically, for my purpose here, Chandralekha, the pioneer of Indian contemporary dance, who sadly passed away on 30 December 2006, did not see tradition and modernity as two separate things, and she believed that Asian modernism did not have to follow Western modernism. She once said, 'I wanted to show how Indian dance could be modern on its own terms without borrowing them from the West' (Chandralekha, 2006: 95). Ramli Ibrahim of Malaysia has also asked questions that follow on from what Chandralekha believed in:

> How relevant is the preoccupation of Euro-American modern dance movement and aesthetics in relation to the Asian psyche? [...] How do we relate and communicate relevant, localized, and immediate issues pertaining to the people, the landscape, the environment, and

the interactive and diverse socio-politico-economic impulse around them? [... W]hy not capitalize on these sources, which were already imprinted in the unconscious of the Asian mind?

(Ramli, 2003: 27)

He goes on to argue: 'It is now possible for [Asian] modern works to be created independent of the Euro-American tradition. There is no need for Asian modern creators to rummage the rubbish cans of European and American experiments.' (ibid.: 29) Contemporary dance practice and choreography as a form of research must engage with its immediate and localized environment, and its past cultural traditions.

An important figure in Indonesian contemporary dance whose work was strongly based on local tradition – here West Sumatran martial arts, or *pencak-silat* – was the late Gusmiati Suid. Characterized by sharp and speedy movements, Gusmiati's choreography was strong, simple, beautiful, and often imbued with sensitive feeling. Two elements are predominant in her work: *adat*, or traditional wisdom, and *syarak*, or Islamic teaching. Choreographic form could be new and adopted from the West, but movements, vocals and music were almost always dominated by Minangkabau artistic nuance; meanwhile, the content of the work remained attached to local wisdom, Islamic teaching, and humanistic values. For more than ten years, Gusmiati Suid and her Gumarang Sakti Dance Theatre were iconic in Indonesian contemporary dance. She passed away in 2001, and her talented dancer-choreographer son, Boi G. Sakti, has continued her mission to nurture the growth of contemporary Indonesian dance. He says, of the company's practice-research approach to dance creation:

> Our company is strongly based on the Minangkabau *pencak silat* tradition. We use the materials, but don't rely on the form. We study the cultural philosophy of self-awareness, spontaneity, strength, and feeling. Therefore, whatever form the new dance takes, we are always in this essential state of awareness. We don't need to look for new movements. We just re-use the essence of the tradition no matter how the movements are sequenced or used.

> (Boi G. Sakti, 2006: 110)

Conclusion

In Indonesia and other Southeast Asian countries, collaboration is one trend in the creation of contemporary dance. Collaboration helped Sam

Sathya – as earlier discussed – to 'work out of the box' and to gain a sense of how dance forms can be developed. But collaboration is, of course, a process not free from pitfalls. As Malaysian choreographer Marion D'Cruz observes:

> Collaborations, at their best, are profound learning experiences that empower the collaborators to expand their visions and their creativity, give them courage, critique and support to move into areas of work that they might otherwise not venture into, and provide exciting spaces for experimentation. At their worst, they allow some collaborators to colonize others and this is a most disempowering experience! However, one strives to do collaborations for potentially profound experience that brings.

> (2003: 77)

Through collaboration, one works and engages with others and, as a consequence, can better understand oneself and one's own culture, a point that Pichet Klunchun emphasizes in *Pichet Klunchun and Myself*. Pichet says to Bel:

> I want to do two things. First, I want to bring young people in Thailand to see and appreciate their tradition. Second, I want to invent my own dance movements that are still connected to Thai tradition. I went to different parts of the world to study different dances: flamingo, American modern dance, contact improvisation, jazz, and the cool hip-hop dance. I enjoyed learning and doing those dances [...]. Then I decided to stop practicing those dances and go back to explore my traditional *khon* dance movements.

Chandralekha described such a process of what is basically cultural self-reinvention as 'an inward journey into one's own self; a journey constantly relating, refining the reality of the in-between area; to enable tradition to flow free in our contemporary life' (cited in Schmidt, 1997: 47). Thus, in Indonesia, about 30 years ago, choreographer Sardono W. Kusumo worked at 'enabl[ing] tradition to flow freely in our contemporary life.' Reflecting upon that time, he writes:

> Being in the Western world allowed me to distance myself from and to analyze any environmental phenomena I had experienced in the past, without feeling empty or defeated, as many psychologists suggested [would happen to] many people [living in the modern world, and

who] are loosing their old values while not yet acquiring the modern values. I have both tradition on my right hand and the future on my left. The very moment is a 'surprising moment' for the dialectic meeting between the two elements will give birth to a new form I never knew before.

(Kusumo, 2004: 29)[7]

For nine months in 2002, Joyce S. Lim, Malaysian dancer-choreographer conducted research on contemporary Indonesian dance. She was specially concerned about finding out the particular elements of tradition a choreographer maintains in developing their choreographies from traditional into contemporary dance:

> The answer [to my concerns], although [taking place] in various forms, always came back to one point – you can change the shape and form of the dance, but *rasa*, *taksu*, or *roh* (depending on whether a Javanese, Balinese, or Sumatran was responding) must always be there. It is difficult to fully describe what *rasa*, *taksu*, or *roh* is [*sic*]. It refers to the inner feeling/spirit/strength of the dancer, of being in total immersion, comprehension and awareness of one's dancing at that moment.

(Lim, 2002: 4)

What Lim argues for reiterates Sardono's point that the contemporary reinforces a 'dialect[ical] meeting between the two elements [of the "old" and the "new"].'

In Thailand, as in Indonesia today, the traditional performing arts belong neither to the kings nor to the people, but to tourists. These are arts packaged following the packaging style of the mass-culture industry, and turned into pure commodity, into entertainment. In the past, the traditional performing arts – dance included – were an integral part of the social, cultural, and religious life of the people. They were performed in various festivals, ceremonies, and religious rituals to strengthen tolerance and bring people together. Today, treated as a commodity and secularized, traditional dances are performed not for its meaning but for money. As such, traditional dances are already uprooted from their soil: local beliefs, rituals, mythos, the sacred, are all in some sense lost. Given this dilemma many Indonesian choreographers have continuously searched for the contemporary in the traditional. By making continuous dialogue with their respected traditions, they create new, contemporary work without sacrificing basic values.

Notes

1. An impressive collaborative work presented at the 2006 IDF VIII was *Heroine*, a solo piece choreographed by German choreographer Arco Renz, and performed by the Taiwanese dancer Su Wen-Chi.
2. Free translation of text into English by Murgiyanto.
3. The performance occurred at the International Symposium on 'Cultural Identities, Artistic Identities: From Bombay to Tokyo,' Centre national de la danse (CND), Pantin/Paris, France, 12–16 January 2006.
4. This performance took place in Taipei, Taiwan, at Novel Hall in June 2006.
5. During the performance in Taipei, I managed to take notes on the dialogue between Pichet Klunchun and Jerome Bel in their *Pichet Klunchun and Myself*. My attendance at two other programs by Bel (his *Nome donné par l'auter* and *Jerome Bel*) that also took place in Taipei between 2 and 3 June 2006, along with an open discussion with the choreographer, increased my understanding of his ideas on conceptual choreography. It is important to note that the texts I quote were performative in nature and not 'actual' academic text.
6. Also see Tagore's statement cited in S. N. Hay: 'I have done what was possible – I have made friends. Some of your patriots were afraid that, carrying from India spiritual contagion, I might weaken your vigorous faith in money and materialism. I assure you those who thus feel nervous that I am entirely inoffensive [...]. I can even assure them that I have not convinced a single skeptic that he has a soul, or that moral beauty has greater value than material power. I am certain that they will forgive me when they know the result' (1970: 184).
7. Free translation of text into English by Murgiyanto.

References

Chandralekha, 'Artists' Voices (Chandralekha in conversation with Urmimala Sarkar Munsi),' in *Shifting Sands: Dance in Asia and the Pacific*, ed. S. Burridge (Canberra: Australian Dance Council, 2006).

D'Cruz, M. 'Collaborative Efforts, Experimentations, and Resolutions,' in *Diversity in Motion*, ed. Mohd. Anis Md. Nor (Kuala Lumpur: MyDance Alliance and Cultural Centre, University of Malaya, 2003), pp. 77–89.

Dibia, I. W. 'Karya Seni Lintas Budaya dan Beberapa Permasalahannya: Body Tjak Sebagai Studi Kasus' ('Intercultural Art Work and Its Problems: *Body Tjak* as a Case Study'), in *Seni Pertunjukan Indonesia: Menimbang Pendekatan Emik Nusantara (Indonesian Performing Arts: Considering the Nusantara Emic Approach)*, ed. Waridi and B. Murtiyoso (Surakarta: Program Pendidikan Pascasarjana and the Ford Foundation, 2005).

Hata, Y. 'Children who Kill their Fathers,' in Japan Foundation Asia Center Program for *Lear*, Teater Tanah Air, Beautiful Miniature of Indonesia Park, Jakarta, Indonesia, 5–7 February 1999.

Hay, S. N. *Asian Ideas of East and West: Tagore and His Critics in Japan, China, and India* (Cambridge, MA: Harvard University Press, 1970).

Hujatnikajennong, Agung. 'Foreword,' Program for *Sunken Sea: Sardono W. Kusumo in Collaboration with Sunaryo*, Selasar Sunaryo Art Space, Bandung, Indonesia, 18 July 2006.

Kusumo, S. W. *Hanuman, Tarzan, Homo Erectus* (Jakarta: Ku/Bu/Ku, 2004).

———. 'The Tsunami Quest,' Program for *Sunken Sea: Sardono W. Kusumo in collaboration with Sunaryo*, Selasar Sunaryo Art Space, Bandung, Indonesia, 18 July 2006.

Lim, J. S. *On the Development of Contemporary Dance in the Philippines and Indonesia*. Unpublished report for the Asian Scholarship Foundation, Cohort III (2002).

Ong K. S. 'Lear: Linking Night and Day,' Japan Foundation Asia Center, Program for *Lear*, Teater Tanah Air, Beautiful Miniature of Indonesia Park, Jakarta, Indonesia, 5–7 February 1999.

Ramli Ibrahim, 'Indigenous Ideas and Contemporary Fusions: The Making of Malaysian Contemporary Modern Dance,' in *Diversity in Motion*, ed. Mohd. Anis Md. Nor (Kuala Lumpur: MyDance Alliance and Cultural Centre, University of Malaya, 2003).

Sakti, B. G. 'Artists' Voices (Boi G. Sakti in conversation with Alex Dea),' in *Shifting Sands: Dance in Asia and the Pacific*, ed. S. Burridge (Canberra: Australian Dance Council, 2006).

Sathya, S. 'Artists' Voices (Sam Sathya in conversation with Kang Rithisal and Amrita Performing Arts),' in *Shifting Sands: Dance in Asia and the Pacific*, ed. S. Burridge (Canberra: Australian Dance Council, 2006).

Schmidt, J. 'The Backbone of Freedom: A Statement by Chandralekha,' *Ballet International/Tanz Aktuell*, 7 (1997).

Sugiharto, B. 'Revitalitas Tradisi' ('Revitalizing Tradition'), in *Seni Pertunjukan Indonesia: Menimbang Pendekatan Emic Nusantara* (*Indonesian Performing Arts: Considering the* Nusantara Emik Approach), ed. Waridi and B. Murtiyoso (Surakarta: Program Pendidikan Pascasarjana and Ford Foundation, 2005).

Sunaryo. 'On the Mountain of Wind,' Program for *Sunken Sea: Sardono W. Kusumo in collaboration with Sunaryo*, Selasar Sunaryo Art Space, Bandung, Indonesia, 18 July 2006.

Tagore, R. *Nationalism* (1917; Delhi: Rupa, 1994).

Waridi and B. Murtiyoso. (eds) *Seni Pertunjukan Indonesia: Menimbang Pendekatan Emik Nusantara* (*Indonesian Performing Arts: Considering the* Nusantara Emik Approach) (Surakarta: Program Pendidikan Pascasarjana and Ford Foundation, 2005).

13
Word and Action in Israeli Performance

Sharon Aronson-Lehavi and Freddie Rokem

Reviving Hebrew and materializing Zionism: word and action converge

Israeli culture coordinates many complex levels, including the actions and cultural expressions constituting a national culture, the language(s) developed to describe these actions through day-to-day reporting, the more distanced writing of history, and the artistic/aesthetic expressions that a national culture produces. These forms of coordination, regardless if they are conscious or not, serve both as an expediting element to achieve certain political, ideological, and cultural-expressive goals, as well as a dynamic mechanism for reflection and critique. One distinguishing feature of early Zionist ideology and its realization in the 1948 establishment of the Israeli state was that the categories of 'doing' and 'speaking/writing' were very closely connected and coordinated, constantly mirroring each other, and expressing from various perspectives the notion of fulfilling a dream: the return of the Jews to their ancient homeland.

The basic and even radical harmony between these two activities – the doing and the speaking – has served as the basic performative unit of Israeli culture. Frequently, this unit created a powerful, hegemonic junction between the specific forms of doing and speaking that both constituted and fulfilled the ideological goals of Zionism, enabling the multi-faceted return of the Jews to their ancient Biblical homeland. This return, fulfilling a very deep sense of longing and even nostalgia, was simultaneously realized *physically* by settling and developing the country and *spiritually/aesthetically* by reviving Hebrew as an everyday, vernacular spoken, performed, and written language. The Hebrew language, which until the end of the nineteenth century existed only as a holy language used within a set of religious actions and practices, was gradually secularized and adopted to the modern world (of nationalism), at the same time as it served as the major expressive form for the revival of a modern secular Hebrew culture, including the theatre. The first performances in Hebrew, in the 1890s (when the country was still under Ottoman rule), were performed in schools, and one of their major aims was to teach Hebrew as an everyday spoken language to schoolchildren as well as their parents. When the Habima theatre, today Israel's National Theatre, was established in Moscow in 1917, a central aim was to develop Hebrew as a spoken language. Similarly, this convergence of word and action through theatre can be seen in the 1944 establishment of the Cameri Theatre of Tel Aviv (today one of Israel's leading repertory theatres), with its explicit emphasis on developing Hebrew written drama that depicts local, topical, and contemporary issues.

We contend that this change in the use and status of the Hebrew language, transforming it from a language of religious practice into the hegemonic ideological backbone of Israeli secular culture – and here the theatre has played an important role – is *the* most significant performative act of Israeli culture, because it marked the difference between a Jewish identity and an Israeli Hebrew-speaking Jewish identity, as well as created a separation between Jewish Hebrew-speaking Israelis, and those Israelis who speak Arabic or other languages as their first language.[1]

The act of speaking Hebrew on a daily and secular basis, alongside practicing the national and Zionist ideology, created for the first time in Jewish history a cultural identity that was paradoxically nurtured from its religious origins (i.e., the Bible, the idea of the Promised Land, and the Hebrew [holy] language itself) and yet, post-Holocaust, in fact rejected through this very act many signifiers that tied it to an image and identity of religious Jewishness.[2] In other words, speaking Hebrew on a daily basis became a dualistic act of simultaneously reconnecting to *and* rejecting a post-Holocaust stereotypical Jewish identity, creating in the modernist sense of the word a 'new *man*,' who is Israeli no less than Jewish. This mutual interconnectedness between action and language aimed to create an Israeli national identity liberated from the humiliation, suffering, and pain of exile, laying the basis for a contemporary and historically relevant Jewish/Israeli culture expressing itself in Hebrew.

This self-reflective mechanism, enabling the Jewish people to supposedly re-enter history – through which the arts were also included in the 'doing' – as the realization of the Zionist project, has been crucial for all the performing arts in Israel, and for the theatre in particular. This junction and combination between the 'doing' and the 'speaking' has no doubt also posed a unique challenge to the notion of performance and the performative within the gradually developing Israeli culture, where the arts have simultaneously been perceived both as an integral aspect of the Zionist project itself as well as a tool for critique and self-reflection.

The conflict: word and action disintegrate

The hopes for renewal in all fields of art, including the theatre, were certainly fulfilled, and during the pre-state period under the British Mandate (1917–48), as well as during the first two decades after the establishment of the Israeli state in 1948, a broad variety of artistic activities were developed where the harmonious coordination between action and language, politics and artistic creativity was not seriously challenged. However, as the state of Israel was reaching its twenty-fifth anniversary in the early 1970s, this close linkage was gradually and more seriously probed, destabilizing the hegemonic structures of Israeli society, and critical discourses in artistic activities as well as through other forms of critical writing, including academic research, were developed. At this time, the 'benefits' of the military victory in the 1967 War, leading to the occupation of the West Bank and the Gaza Strip, inhabited by a gradually growing Palestinian population, were perceived as more and more morally compromising. And the combined surprise attack of the Syrian and Egyptian armed forces in the 1973 October War demonstrated the vulnerability of Israel. These events – followed by the war in Lebanon (beginning in 1982), the Palestinian uprisings (the first Intifada beginning in 1987 and the second Intifada beginning in 2000, separated by the establishment of the Palestinian Authority in 1993 as a result of the Oslo Accords), and most recently the war in Lebanon in 2006 – gradually have clarified the cultural and moral complexities Israel is facing.

In addition to 'word and action' as a theoretical tool to analyze Israel's cultural performativity, we would like to suggest three main signification systems that have largely shaped Israeli identity and its culture. The relationships and, in particular, the tensions between these signification systems have become intensified after the 1967 War:

1. The Bible (Old Testament), which functions as the local and national mythology and the original bearer of the Hebrew language, and which, together with European anti-Semitism, is the *raison d'être* of the Zionist project.

2. The Holocaust, which lives on as a national trauma of victimization with almost mythical dimensions and as a historical event that led to the foundation of the State of Israel.
3. The ongoing state of armed conflict, especially with the Palestinians, through which the two peoples are desperately entangled with each other, unable to resolve the conflict.

Whereas Israel's wars with its neighbor countries have been mostly considered necessary for defense of the state, the Israeli-Palestinian conflict is not only characterized by a lack of consensus within Israeli society regarding the occupation and its moral consequences, but it has also become an absurdist and existentialist situation of mutual retaliation. In the following pages we will examine a few plays and theatre performances that can be categorized as avant-garde and experimental in both form and content and that demonstrate changes in the basic performative unit of 'word and action.' In addition, the performances we look at address in different ways the three signifying systems mentioned above (Bible/religious iconography; Holocaust; and Israeli-Palestinian conflict).

The existentialist coupling of victim/victimizer has become a central dramatic and theatrical mechanism of the *œuvre* of the late Hanoch Levin (d. 1999), Israel's most significant playwright and director to date. In addition, the convergence of the Bible, Holocaust, and Israeli-Arab conflict can be found in many of his plays. Levin published more than 50 plays during his lifetime, out of which about half were performed on stage, mostly directed by Levin himself. In many of these plays and performances, he exposed the complex relations of mutual humiliation between victimizers and victims: family-members, mythic figures, or players in an imagined political arena.

Levin's play *Murder* (performed at the Cameri Theatre of Tel Aviv in 1997, directed by Omri Nitzan) is probably the play most firmly anchored in a concrete political reality, even though it contains many 'fantastic' elements. It begins with a scene where a group of Israeli soldiers are beating a Palestinian youth to death. When the young boy's father asks the soldiers how he died, they say they do not know because he was already dead when they arrived. The next scene shows how several years later the Palestinian father takes revenge on a newly married couple celebrating their first love on the city beach, killing first the groom (whom he believes is one of the soldiers who killed his son) and then, after raping her, also the bride. The third scene shows an angry Israeli mob lynching two Palestinian workers caught peeping into the apartments of the neighbourhood where they are employed. The play, as well as the production, forcefully and even cruelly reflects and complements the kinds of information and forms of presentation that the Israeli public constantly receives about the conflict through media and eyewitness reports.

Levin's plays are based on two simultaneous narrative strategies, partly overlapping but at the same time also in conflict. The first is based on constructing a loose, but seemingly inevitable causal relationship between the play's major events. This reflects the 'deterministic' aspect of the conflict's chain of events, wherein the historical or cultural master-narrative (typically that of a political-ideological conflict), is insoluble. This chain of events creates its own intrinsic and vicious logic, where the violent events are more or less inevitable. The second narrative strategy shows how seemingly illogical events, nonetheless, take place without any prior warning. Things just happen, even if they have been described as 'impossible' or 'unthinkable.' And when they do take place, they are actually even more violent and cruel than can be imagined or articulated. This Artaudian strategy, in part reflecting the post-Second World War crisis of representation, could loosely be termed a 'narrative of terror,' because terrorism occurs without prior warning. Together, this complex combination of narrative strategies potently reflects a post-Holocaust, collective Jewish-Israeli historical consciousness, at the same time as it signals a crisis in the relationship between word and action.

Although Levin's dramaturgy evokes absurdist situations in which any action cannot but lead to a worse situation, his ironic, poetic, and richly layered language investigates the disintegration of the defining performative unit of Israeli culture: language as action. Whereas his rich language attests to the potency of artistic expression, his dramatic situations expose in a Beckettian manner the hollowness of action. For example, in his 1998 play *Those Who Walk in the Dark* (*Ha-Holchim ba Hoshech*), the play's title is itself an ironic quotation taken from Isaiah's prophesy: 'The people walking in darkness have seen a great light' (*Isaiah* 9:2). Rather than seeing a great light, the play literalizes and fragments the characters' most internal thoughts, to the point that meaning is lost. At a certain point, when God appears on stage, he is asked why he created evil. He gives an answer, but at that very moment the loud noise of a passing train makes it impossible to hear. The train itself clearly stands for the Holocaust, once again pointing at the inability of language to express meaning.

Identity politics and performance: word and action are deconstructed

International performance trends, displacing traditional theatre's role as an expression of avant-garde culture, have had a profound influence on Israeli performance and performance art. Feminist and postcolonial paradigms of expression have gradually enabled non-hegemonic bodies to voice themselves. This has further deconstructed the relations between word and action and enabled the re-examination of Israeli identity by exploring the performer's culturally multi-layered body. In this section we examine two performances in which the body of the performers becomes a site through which the signifying systems filter and are negotiated: Rina Yerushalmi's *Bible Project* (1995–2000) and Semadar Ya'aron's *Wish upon a Star* (2005).

Both performances depend on a direct and literal theatricalization of national and mythical texts and symbols in order to problematize and even subvert the Israeli mechanisms of inclusion, exclusion, and representation. Whereas Yerushalmi's epic performance was performed by an ensemble of 12 actors, Ya'aron's work is a solo performance art piece. Yerushalmi's performance includes many themes, among them feminist issues, but even when the scenes are not about women, her innovative interpretations of the biblical excerpts express the option of reappropriating traditional signifying systems from unexpected viewpoints. The feminist entrance into political and mythological zones traditionally dominated by patriarchal cultural structures demands a double performative act: first, a woman has to make place for her own body and voice within these texts and narratives, and second, only then can she seek to re-examine the traditional viewpoint and evoke a multi-vocal and complex identity.

Israeli theatre history is marked by numerous dramatic adaptations of biblical stories and theatrical 'confrontations' with the Bible itself, precisely because of its linguistic, historical, and mythical status and its powerful presence in the collective Israeli consciousness. Rina Yerushalmi's eight-hour-long Bible project, arguably one of Israel's most significant theatrical achievements, consists of two independent, but interrelated productions: *And He Said. And He Walked* (*Va Yomer. Va Yelech*), which premiered in 1995, and *And They Bowed. And He Saw* (*Va Yishtachu. Va Yare*), which premiered in 1998. Both productions were performed for several years in Israel and at festivals throughout the world. The entire performance is based on over 80 original texts from the Hebrew Bible, selected, edited, and directed by Rina Yerushalmi, and performed by the Itim Ensemble, which Yerushalmi founded in 1989.[3] The performances were structured as a collage of biblical excerpts presenting a complex and provocative vision of Israeli society. By keeping the texts in the original archaic and poetic language and in the third person ('And he said'), Yerushalmi's actors avoided conventional enactment of characters, instead letting familiar texts and stories filter through their moving bodies and voices. The contrast between the performances' modern and secular look – an almost empty, brightly lit space with 12 actors in modern black clothes – and the texts' antique and highly charged religious status enabled the use of irony, creating Brechtian effects of defamiliarization that revealed new meanings in these texts, while also deconstructing the traditional site of biblical reading and recitation and thus allowing traditionally silenced voices to be heard.

Especially topical was the second part of the first production, *Va Yelech* (*And He Walked*), which draws parallels between the biblical story of the Exodus from Egypt and the conquest of the Land of Israel and the twentieth-century ordeals of the Jewish people. The 12 actors open this part holding brown suitcases that in Israeli culture have become icons of the Holocaust and thus of signifying Holocaust survivors. Gradually, however, the actors become transformed from a group of victimized wanderers into a solid, martial group ready to conquer the Land of Israel, fight in devastating wars, and even sacrifice their children. Such willingness has been narrated many times in Israeli performances through variations of the story of Abraham and Isaac (*Genesis* 22).

This part of Yerushalmi's performance culminates in the story of the Daughter of Jephthah (*Judges* 11, 29–40), which, similarly to the Greek myth of Iphigenia, tells about a father, Jephthah, the judge (leader/warrior), sacrificing his daughter because of his oath to God to sacrifice the first to welcome him when he returns victorious from the war. When he arrives home, his young daughter runs towards him 'with timbrels and with dances.' Obligated to keep his promise, Jephthah sacrifices his own child. Whereas traditional interpretations either usually stress the necessity of human sacrifices as part of 'building a nation,' or Jephthah's tragic fate for having to kill his daughter (who, incidentally, is only referred to as 'Jephthah's daughter'), Yerushalmi instead creates a ruthless and powerful *mise-en-scène*. In this particular scene, the performer Iyar Wolpe repeats the biblical text verbatim five times, without changing a word, as her body movement gradually becomes more and more intensified, accompanied by increasing loud electronic music. Standing alone in the middle of the performance space, hardly moving from the fixed spot from which she 'recites' the biblical text, she transforms from a storyteller into a performer in trance.

In the preceding scene, the other actors had just completed 'conquering' the Land of Israel by symbolically covering the performance space in a geometrical pattern with white pages bearing biblical verbs, indeed tying word to action. The floor thus simultaneously presents an image of the conquered land which also looks like a military cemetery, in the middle of which Wolpe/the daughter makes herself a small space by moving one of the pages, as if marking her own grave. It is from this spot that she puts on her electrifying performance-within-a-performance. Rather than the traditional paternal viewpoint, the presence and perspective of the condemned daughter becomes immediately evident.

However, finding and voicing the silenced young girl is only the scene's first step. By changing her intonations and emphasizing different gestures each time, Wolpe switches her perspective from a detached storyteller into a young and terrified girl, then into a raging woman, and then into the agonizing father about to kill his own daughter. By repeating the long text five times through these shifting identities, Wolpe's body voices the victim and yet, at the same time, escapes becoming a victim herself. She offers the audience a multi-perspective vision of suffering, pointing out its absurdity from all sides, using the ritualistic repetition and trance to beg for an end to her pain while at the same time marking its endlessness. At the very end of this scene, Wolpe breaks out with the cry 'A-vi!' – 'my father!,' creating a shattering moment of personal presence and distress after having totally dissected the biblical text without changing a word.

Following this scene, *Va Yelech* ends with God's command to Abraham 'Lech-Lecha' ('Leave your country, your people and your father's household and go to the land I will show you. [...]' *Genesis* 12:1). This command – the mythical unit that designates the Land of Israel to the Jewish people – incorporates both the idea of salvation (*geula*) and that of wandering and exile (*galut*). Therefore, Yerushalmi's choice to end the performance with this command marks this cyclical pattern not only as part of history but also of the present.

A Wish upon a Star, which premiered in 2005, is a solo performance by actress and avant-garde performance artist Semadar Ya'aron, known for her extreme and total acting methods, as well as for her deep commitment to political change. In 2009 she became artistic director (with Moni Yosef) of the Acre Festival for Alternative Theatre, where in 1991, she and former partner Dudu Ma'ayan premiered one of the most shocking and penetrating performances in Israeli theatre history, *Arbeit Macht Frei vom*

Toitland Europa (*Work Liberates from Death-Land Europe*). Its title incorporating the text from entrance signs to Nazi death camps, this performance examined Israeli culture through the lens of the Shoah's impact on Israeli society. Some of the issues explored in the performance included a critique of memorial Holocaust ceremonies in the Israeli education system, growing up in Israel as a 'second generation' of Holocaust survivors, and Israeli attitudes toward Arabs as the Other. Displacing the Nazis' cynical use of the slogan '*Arbeit Macht Frei*,' this performance instead suggested that by doing this 'work' in the theatre and confronting the complex interaction between the past and the present, Israeli society can perhaps free itself from its tragic past in the Death-land of Europe. But even if this performance pointed out the direction where Israeli society might free itself from past traumas, it ended in a dystopian vision where basic human values were disregarded.

For that performance, Semadar Ya'aron experienced hunger by starving herself and tattooed the date of her father's death on her left arm, as if she were a Holocaust survivor. In *A Wish upon a Star*, Ya'aron posits herself vis-à-vis her national identity. The performance opens with a woman, Ya'aron herself, proposing marriage to a Star of David, one of the most important Jewish religious and national symbols (also used by the Nazis to mark the Jews), which she 'marries' in order to be impregnated with the new Messiah. The Star itself is an extremely heavy metal, human-sized construct, wrapped with shrouds and a Jewish prayer shawl (*talith*) hanging on one side to mark its masculinity. Traditionally, Jews are also buried in their prayer shawl. The performance depicts Ya'aron's imaginary journey to Jerusalem with the Star, highlighting her interaction with Jewish and Israeli nationality. Ya'aron hangs herself acrobatically on the Star, reaching the height of 12 meters without a safety net on the ground; at another point, she lies down on the floor as the Star (controlled by an engine and a remote control) sacrilegiously penetrates her, accompanied by the music of Richard Wagner. And it must be noted in this context that until recently it was taboo to play Wagner's music publicly in Israel because of his ideological influence on the Nazis.

During the performance, Ya'aron also switches identities and situations by telling her most personal memories and life experiences on the one hand and addressing national issues on the other. For example, although she marries the star in order to give birth to the Messiah, she ends up confessing that it is hard for her to play a pregnant woman since she was and never will be pregnant, intimately tying together her personal story with the 'big' story. Similarly, Ya'aron turns her own body into the map of Israel, on which she marks cities and villages that used to have Arab names but now have Hebrew names. Using her own body as a site for social critique is a performative act of taking responsibility on the most personal level. This, ironically, contrasts the bond between her very body and the national symbols she interacts with, creating a paradoxical situation that is based on criticism and reinscription of these symbols and narratives.

Throughout the performance Ya'aron ruthlessly rips open every value on which Israelis have been brought up and does not avoid problematizing the complexity of the Israeli identity. And yet, at the end of the performance, Ya'aron quietly and tiredly carries the huge and heavy Star of David off the stage, much like Jesus carrying the cross, a cultural heritage one is bound to. The bombastic inflation of cultural symbols that Ya'aron performs and deconstructs creates an overwhelming and disorienting experience.

In a culture long dominated by patriarchal, militaristic, and nationalistic discourses, the performances of Yerushalmi and Ya'aron confront the myth in a Grotowskian sense. By locating performance in the body, they evoke images of a multi-layered and complex cultural identity and use the theatre as a ritualistic space.

Prospects for the Future I: word and action reconsidered

The impressive modern renaissance of the Hebrew language has produced a mono-linguistic culture. There have, however, been several attempts to confront and question Hebrew's hegemonic position in Israeli culture and in the theatre in particular, especially under the auspices of the annual Acre Festival for Alternative Theatre. Such attempts offer one way to reconsider the cultural structures established by Zionism. During the last decade, several Israeli films have also represented a multilingual reality. This development has not yet fully reached the theatres, but will in all likelihood become more and more frequent in the future, encouraging the audiences to re-evaluate the complex relation between word and action.

One interesting example of diglossia is the production of Beckett's *Waiting for Godot*, directed by Ilan Ronen (today the artistic director of the Habima theatre), at the Haifa Municipal Theatre in 1985. In this production, Didi and Gogo were played by Israeli-Arabic actors. When the performance played before Arabic audiences, Didi and Gogo spoke in Arabic, and when the play was performed for predominantly Jewish audiences, they spoke in Hebrew with a marked Arabic accent. Pozzo always spoke in Hebrew, while Lucky spoke in a distorted Classic Arabic.

The set, instead of the tree prescribed in the original stage directions, showed a pole from a building site, transforming the play's abstract universal location into a concrete Israeli situation, where Didi and Gogo have become Palestinian construction workers waiting for a solution to their situation.

Since the two Palestinian *Intifadas* and the terror attacks on Israeli cities, Palestinians are hardly employed in Israel, and instead workers for construction and the like are brought from third-world countries. Therefore, Ronen's political interpretation of Beckett's play and its location on a construction site would be considered dated nowadays. However, the 'glocal' phenomenon of migrant workers as a group of deprived people is another social problem Israel is facing, and one that brings up questions of multiculturalism. *The Second Law of Thermodynamics* (2004) is a work by performance artists Tamar Raban and Guy Gutman with Ensemble 209, which evokes an image of a multicultural society. The location of Raban's performance space in the midst of Tel Aviv's huge bus station, a very poor area where communities of migrant workers live, gave inspiration to the performance that stages lessons from education systems that are designed to unify people. As the lessons are 'taught,' the performers share with the audience personal recollections of social exclusion. The performers include an Arab Israeli performer, a Jewish Israeli performance artist, and newcomers to Israel from Russia, Ethiopia and Romania. Performing together, they create a Babylonian sense of difference and evocation of openness on the one hand, whilst demonstrating the similarities between their experiences on the other.

When the big wave of Jewish immigrants from Russia arrived in the early 1990s – all in all more than one million, constituting almost 20 percent of the Israeli population – they kept Russian as their language and even founded their own theatres. The most successful is the Gesher Theatre (*gesher* means 'bridge' in Hebrew), founded in 1991 by an immigrant group led by director Yevgeny Arye. They began by playing in Russian but gradually began to play all of their performances in Hebrew. The first production at the Gesher Theatre was Tom Stoppard's *Rosenkrantz and Guildenstern are Dead*, which was initially performed in Russian and only later 'translated' into Hebrew, without the actors really understanding this language at all. This production both reflected and exposed the situation of the newly founded company, showing how a group of actors arrives at a castle in a strange country without understanding why they had been invited or what they were supposed to do. The Gesher actors who had supposedly 'returned' from exile to their ('ancient') homeland presented a group of completely disoriented actors who had arrived in

Elsinore to stage a performance. Since this first production, however, several Israeli-born actors have joined the group, and in 2001, for the first time, the theatre mounted a production (of Strindberg's *Miss Julie*) that was not directed by Arye himself, but by an Israeli-born director, Yossi Yizraeli. Throughout its existence, the Gesher Theatre has constantly reformulated its own identity in terms of the dialectical tensions between exile and home, poignantly echoing one of the major ideological and cultural concerns of Israeli society, a society based on a founding myth of the ingathering of the exiles.

Since the Gesher Theatre was founded, the group has undergone a gradual assimilation process into the hegemonic Israeli theatre culture, while at the same time discarding some of the 'Russian' characteristics of its founders, even if the Russian accent of its original actors is still quite prominent. However, in a country where the national theatre, the Habima theatre, was founded in Moscow on the basis of the Stanislavski-Vakhtangov traditions, and where Jurij Lubimov, another prominent Russian émigré, also worked for several years in the 1980s, the sources of the hegemonic aesthetic norms and the forms of artistic authority they represent are quite complex. The Russian modernistic theatre traditions serve as the dominant professional and stylistic sub-text for the Israeli theatre, even if at various points, but never to the same extent as the Russian traditions, this stylistic and professional foundation has also been shaped by German and Anglo-Saxon influences.

The dialectical tensions between the Diaspora and the idea of homecoming are for Israeli society and culture in general, and for the Israeli theatre in particular, not only an ideological concern, but an aesthetic one as well, constantly redefining and even challenging the sources of its own short history and heritage. The Gesher Theatre, with its distinct acting style and its grand stylized theatrical gestures, paradoxically represents a return to the 'foreign' birthplace of professional Israeli theatre, and at the same time it must be considered an example of absorption and assimilation to the new country. The Gesher Theatre obviously constitutes a cultural bridge that leads in two directions. At the time this chapter is being written, there are also Israeli theatres performing in Ethiopian, in the Jewish Moroccan dialect, as well as in Yiddish and Ladino.

Prospects for the Future II: the body's expressiveness

The complexities of contemporary Israeli identity – with its constant and violent shifts between a traumatic history and an ongoing existential insecurity and state of conflict – have been deconstructed and unmarked as well as reconstructed both in performance and critical discourse. Israeli performance culture has become a site where coded systems can be confronted and played out against each other, opening up a wide range of human and narrative possibilities. In the arena of avant-garde art and performance, it is possible to distinguish a gradually growing suspicion against ideological languages and meta-narratives.

Beginning in the early 1990s, one interesting result of this suspicion has been the growing reliance on the human body's direct expressiveness in performance. The body has a different presence than words, since it is the action itself, not an additional layer of commentary and reflection. This recognition of the body's direct expressiveness has led to the remarkable development of dance-theatre and movement performances during recent years. The Batsheva Dance Company, led by dancer and choreographer Ohad Naharin, is perhaps the most famous company internationally, but there are many other local companies and artists whose work is created within an international context and who have large, enthusiastic, and young local audiences. Yasmin Goder Dance Ensemble is one such company. It is still too early, however, to say how this development will affect the larger scene of Israeli theatre and performance culture, but it holds the opportunity to embrace even more multicultural and multi-linguistic members of contemporary Israeli society.

Notes

1. It should be noted that the official languages of the State of Israel are Hebrew and Arabic.
2. See, for example, Almog, 2000.
3. For more information on this performance and on Rina Yerushalmi, winner of the Israel Prize 2008 for Theatre Creation, see: http://www.itimtheatre.com.

References

Almog, O. *The Sabra: The Creation of the New Jew*, trans. Haim Watzman (Berkeley: University of California Press, 2000).

Aronson-Lehavi, S. (ed.) *Wanderers and Other Israeli Plays* (New York: Seagull Press, 2009).

Assaph: Studies in the Theatre (Tel Aviv: Tel Aviv University, 1985–) (In English).

Avigal, S. 'Patterns and Trends in Israeli Drama and Theater, 1948 to the Present,' in *Theater in Israel*, ed. L. Ben Zvi (Ann Arbor: University of Michigan Press), pp. 9–50.

Ben-Zvi, L. (ed.) *Theater in Israel* (Ann Arbor: University of Michigan Press, 1996).

Brown, E. *Allegory and Irony in the Satirical World of Hanoch Levin* (Ann Arbor: UMI Research Press, 1989).

Kushner, T. and A. Solomon (eds) *Wrestling with Zion: Progressive Jewish American Responses to the Israeli-Palestinian Conflict* (New York: Grove Press, 2003).

Levin, H. *Labor of Love: Selected Plays*, trans. B. Harshav, with an introduction by F. Rokem (California: Stanford University Press, 2003).

Levy, S. *The Bible as Theatre* (Portland, OR: Sussex Academic Press, 2002).

Rokem, F. 'Hebrew Theater from 1889 to 1948,' in *Theater in Israel*, ed. L. Ben Zvi (Ann Arbor: University of Michigan Press), pp. 51–84.

———. 'Refractions of the Shoah on Israeli Stages,' in *Performing History: Theatrical Representations of the Past in Contemporary Theatre* (Iowa: University of Iowa Press, 2000), pp. 27–98.

Snir, R. 'Palestinian Theatre: Historical Development and Contemporary Distinctive Identity,' *Contemporary Drama Review*, 3:2 (1995) 29–73.

Teatron (Theatre): An Israeli Quarterly for Contemporary Theatre (General Editors: Gad Kaynar and Haim Nagid. 1998–) (In Hebrew).

Teoria ve Bikoret (Theory and Criticism) (Jerusalem: Van Lear Institute, 1991–) (In Hebrew).

Urian, D. *The Arab in Israeli Drama and Theatre*, trans. N. Paz (Amsterdam: Harwood Academic Publishers, 1997).

———. *Judaic Nature of Israeli Theatre: A Search of Identity* (London: Routledge Harwood Contemporary Theatre Studies, 2000).

14
Democratic Actors and Post-Apartheid Drama: Contesting Performance in Contemporary South Africa

Loren Kruger

In anti-apartheid South Africa, performance culture was dominated by one institution: the Market Theatre. Opening in June 1976, just before the Soweto uprising, the Market provided a venue for plays that expressed the urgency of the anti-apartheid struggle. While other, shorter-lived theatres – from the Space in Cape Town (1972–81) to black institutions in or near Johannesburg such as FUBA (Federated Union of Black Arts) or Soyikwa (honoring playwright Wole Soyinka) – staged critical drama, it was the Market that provided space, support, and marketing for one of apartheid's most visible exports: anti-apartheid protest theatre. Although some lamented the packaging of protest as 'theatre for export' in plays like *Sarafina* (1986) that flattered overseas spectators' sense of their own political correctness (see Mofokeng, 1996), all agree that at its best, in plays from *Survival* (1976), to *Woza Albert* (1981), to *Born in the RSA*, and *Have you seen Zandile?* (both 1986), anti-apartheid theatre bore witness to both the atrocities of, and the resistance to, apartheid. As the primary venue for this genre of performance, the Market offered not only a stage but also a safe haven for performers whose work and whose very selves might have otherwise been banned or harassed by the apartheid state.

At present, in the second decade after the official end of apartheid in 1994, theatre has to take place in an altered landscape dominated by a profit-driven leisure industry. Particularly in greater Johannesburg, Southern Africa's largest and wealthiest conurbation, a long-term drift of wealth to the suburbs, caused as much by movement to more profitable investment opportunities as by capital flight in the wake of political unrest in the 1970s (Beavon, 2004: 147–94), has led to the hollowing out of the former central business district (CBD), and encouraged the development of self-enclosed commercial, residential, and leisure spaces that urbanist Lindsay Bremner has called 'theme park cities' (Bremner, 2004: 118–37).[1] Driven by local conglomerates as well as transnational capital, this industry is represented above all by casinos. For instance, Gold Reef City, near now-defunct gold mines south of the CBD, includes a long-running show called *African Footprints* among other ethnic flavored tourist kitsch. This program alone does not fulfill the casino's mandated requirement to provide fare of cultural and educational significance; that obligation is met by Gold Reef City's subsidy of the Apartheid Museum across the road. While the museum offers no regular live performances, its layout and program, including separate 'white' and 'non-white' entrances, darkened halls, and loud, even overwhelming audio-visual presentation of apartheid oppression, compels visitors to revise their habitual performance as passive consumers in favor of interaction with – or active withdrawal from – the sights and sounds that clamor for attention. In contrast to the bifurcation of commerce and education marked by the thoroughfare that separates Gold Reef City and the Apartheid Museum south of the former CBD, the Constitution Hill development integrates on a central site, under the auspices of state/capital partnerships, the Old Fort (formerly a colonial and later an apartheid prison, now a museum), the new Constitutional Court and, in preparation, commercial and residential spaces.[2]

If the new complexes attempt profitably to combine commerce and education and compete with theatre's offerings of live performance, the institutions of the media compete by creating absorbing and sometimes sharply critical dramatic fictions, and by luring theatre actors, directors, and writers with prospects of larger budgets and wider audiences. As the passing of apartheid has robbed the stage not only of its powerful enemy but also its chief source of drama and of funding from international agencies that supported anti-apartheid culture, theatre in South Africa is governed once again by Baumol's law, according to which predictably escalating technical, wage, and administrative costs increasingly escape the reach of unpredictable box office and subsidy revenues (Baumol and Bowen, 1966). Theatres not covered by 'national institution' grants (like the Market) struggle to compete with National Arts Council (NAC) funding for music, opera, literature, crafts, and visual arts, dance, and 'multi-disciplinary practice' (NAC, 2005).[3] By contrast, film and television, while more expensive to produce, have better access to more sources of funding. These include government subsidy for cultural industries from the departments of trade and tourism as well as the National Film and Video Foundation (DACST, 1998; NFVF, 2005), and support from commercial ventures, such as the conglomerate Primedia or the cable production company MNET, which produce feature films as well as educational series to meet local content quotas – 33 percent for private broadcasters as against 55 percent for the national broadcaster, South African Broadcasting Corporation (SABC).[4] This combination of public local content mandates and private capital sources has encouraged many former anti-apartheid theatre people to work in part for television, and has been a greater draw than the theatre for younger fiction writers like Sello Duiker, who, before his untimely suicide in 2004, was contributing writer for SABC-Education's edgy series about sex, glamour, and AIDS in the city, *Gaz'lam*.[5]

Theatrical exceptionalism

In order to analyze the aesthetic and institutional implications of this new configuration of contesting performance sites, we have to break with 'theatrical exceptionalism.' This phenomenon may seem quaint in technologically developed contexts of performance, but its persistence in African theatre studies still blocks productive research of the forms and institutions of performance in multiple media. By theatrical exceptionalism I mean the focus on theatre and allegedly authentic live performance to the exclusion of hybrid performances involving electronic media and other institutions. A recent, purportedly authoritative collection, *History of African Theatre* (Banham et al., 2004), looks backward to ancient rituals and sideways to informal entertainments but fails, with few exceptions, to look forward to future, and indeed *actual*, cross-fertilization of theatre and media, whether electronic recording and diffusion or mixed-media interpellation of visitors in interactive museum encounters. Theatrical exceptionalism has been reinforced not only by theatre specialists but also by performance studies advocates such as Richard Schechner (1998) or Peggy Phelan (1993), whose conception of performance rests on liveness as the site of authentic expression, even of resistance to the alleged corruption of mediation. As Philip Auslander (1999) has retorted, liveness is not an unmediated category but has become, in theatre's marketing of its difference from media production, a commodity as mediated by ideology as appeals to authenticity.

Two other aspects of theatrical exceptionalism invite critique. The first is the assumption that innovative form necessarily carries a progressive political charge, the second that the theatre's political force depends on its firm separation from capitalist relations of production. Although written in the early days of mass-mediated challenges to theatre, Bertolt Brecht's critique of these assumptions is as relevant as ever.

In 'Primat des Apparats' (1928), Brecht cautions against premature celebrations of 'revolutionary' dramaturgy, in particular plays on topical political issues or stage technology that overwhelms the action (Brecht, 2002: 67). Changing dramatic form or the technical apparatus does not automatically transform cultural institutions or the larger society. Translating *Apparat* as *institution* rather than the habitual 'apparatus,' we can better appreciate Brecht's argument that only a full transformation of the social relations that determine who produces, finances, and appreciates drama – whether in theatre, film, or, in our era, television and other media – should be called revolutionary.

In the *Threepenny Lawsuit* (in Brecht, 2000), published after his unsuccessful suit protesting unauthorized changes to the *Threepenny Opera*, Brecht debunks simplistic oppositions between commodity and art. Instead of the well-worn divide between high (and allegedly commodity-free) art and low (and utterly commodified) mass culture, a distinction which he sees as classically bourgeois in its attribution of an anti-social attitude to art, Brecht argues that capitalist relations of production expose the myth of artistic autonomy and thus challenge culture producers to explode dominant institutions from within as well as without. Refunctioning institutions such as the national broadcaster or prestige stages requires dialectical understanding of the uses as well as abuses of capitalist modes of production (Brecht, 2000: 192–9). Brecht highlights this possibility precisely at the moment when he appears to attack the commodification of his own play as film. He notes that the idea that film is a commodity is something 'we all can agree on,' but he goes on to argue for the 'reshaping force [*ummodelnden Kraft*]' of the commodity form and thus, following Marx if not all Marxists, for the 'revolutionary' moment of capitalism's capacity not only for exploitation but also for transformation [*Verwandlung*] (ibid.: 169).

This formulation of critique *through* the commodity invites us to rethink the still powerful influence of Brechtian theatre of instruction in South Africa, and in particular the use of *Verfremdung* as a key measure of distance from commodified or 'culinary' culture. Usually translated as alienation, *Verfremdung* in fact means *critical estrangement* and thus the *opposite* of alienation. *Dis-illusion*, the first English translation (1936) (Brecht, 1998: vol. 22, p. 960), highlights the process of critique from within rather than at a distance from the commodity form. Brecht's provocative formulation of critique by stealth contrasts with the 'heroic realism' that speaks militant truth. In 'Five Difficulties in Writing the Truth' (1938), he suggests that the 'courage' needed to grasp art as a weapon in time of class war might have to yield in more ambiguous times to the 'cunning' needed to write the truth (Brecht, 2002: 148). This turn to stealth applies not only to cultural critique but, also, as urbanist AbdouMaliq Simone suggests, to the skills that denizens of African cities must acquire to negotiate the urban environment (Simone, 2002: 51).

In cultural production, the tension between heroic realism and critique by cunning shows the difference between the heroic tone of resistance that still marks community theatre, including work on AIDS education, and the cunning mode of critique under cover of commodity production which might characterize public/private museum practice that encourages tourist consumption as well as edification, or of television series that address viewers as agents of their own education *and* as consumers of sponsors' products. Since advertising in discrete spots as well as product placement within the fictional narrative clearly shapes performance aesthetics as well as institutional structure, the institutional context of post-apartheid cultural production should therefore frame any claims about the radical import of formal innovations, from jumpy credits and dissonant music through estranged acting to what Brecht called the separation of the elements, using acting, lighting, music, and so on, to project dissonant rather than harmonious meanings.

This reorientation of performance towards multi-media institutions may disturb those in community theatre who insist on the authenticity, innovation, and educational impact of face-to-face grassroots theatre in combating national crises such as the AIDS epidemic. AIDS education has indeed fueled the work of a wide range of groups from lay activists using the same agit-prop tools as the anti-apartheid theatre like Victory Sonqoba Theatre (Marlin-Curiel, 2004), to medical/artistic partnerships – from small-scale hospital skits (Kruger, 1999: 207–9) to the edu-taining soap opera, graphic story, and radio series *Soul City* (Kruger, 2004), to long-term projects such as *DramAide*, which began with theatre instructor Lynn Dalrymple at the University of KwaZulu in 1991 and became a national program sponsored by the Health Department in 1998 (Dalrymple, 2005). Although these groups claim that their dramatized confrontations between sexually active young people have 'changed attitudes towards HIV,' they concede that sexual practice has not changed to match apparently new attitudes (Bourgault, 2003: 68–74) in this country with more people living with AIDS than any other (five million out of 47 million; HSRC, 2005), a 30 percent infection rate among the most vulnerable group, sexually active women in their twenties, and a former president (Thabo Mbeki) who condoned 'AIDS denialism.'[6]

The acknowledgment that didactic theatre and a behaviorist view of culture have limited impact may be grudging, but it nonetheless opens up critical space for education by stealth rather than heroic emphasis. Education through stealth allows for an exploration of the ways in which mass-mediated performances may contribute, not to an immediately measurable behavior modification, but rather to the more complex creation of what we might call democratic actors. By *democratic actors* I mean South Africans in a range of roles, whether formally defined performers in dramatic fictions, or informally extended to include roles to which targeted audiences might aspire in direct, or more likely indirect, response and to herald the emergence of a democratic South African citizenry beyond the line still marked by the hyphen between post and apartheid.

While this short chapter cannot pretend to present a full study of performance in South Africa, it can offer snapshots of the aspirations and actuality of performance in key institutions in which democratic agency and location are staged and contested. I am particularly interested in the increasing intersection of television and theatre, and the ways in which the new multi-media conditions of production shape the form, content, and personnel of this intersection. I will be looking in particular at the 'edu-taining' but edgy urban series *Gaz'lam* and the recent program at the Market and other theatres, especially the plays of satirist Mike van Graan. Although comparisons with other regions are certainly apt, this discussion focuses on Johannesburg as one of the fasted growing cities worldwide and the exemplary site of economic, social, and cultural experiment as well as risk in contemporary South Africa.

Post-apartheid agency

There is still a gap between city managers' promotion to investors of Johannesburg as a 24/7 environment to match New York and their acknowledgment to residents that crime continues to demand attention and resources (City of Johannesburg, 2002a vs. 2002b; Fraser, 2005).[7] But even if the city's plans for urban renewal are still in process, a new urban culture driven by young, upwardly aspirational if not always fully mobile blacks has found expression in a variety of forms, from clothes – as in the labels Stone Cherry and Loxion Kulcha – or music – where *kwaito*, a house and techno mix, competes with retro sampling of *mbaqanga* and jazz – to television, especially in edu-taining hybrids like *Yizo, Yizo* (which in three seasons from 1999–2002 followed a cohort of students from their last year in a Soweto school to their struggles with university, work, or unemployment thereafter) or, initially by the same producer and with some of the same actors, *Gaz'Lam* (My Blood, My People; 2002–06), as well as in high-end series not specifically labeled as educational, such as the controversial gay content mini-series *After Nine* (2007), which has educated viewers or even determined non-viewers through both positive and adverse publicity as well as the actual broadcast. While the most spectacular city improvement districts (CIDS) have seen mixed use conversion of historic sites such as Constitution Hill and Newtown, both of which combine historic structures (the Old Fort or the Market) with new public as well as private buildings, these new shows highlight less polished urban sites. *Gaz'lam* foregrounds the actual messy location of Johannesburg's cosmopolitan, edgy modernity, the formerly bohemian, now pan-African neighborhood of Yeoville. Here low-rise density now houses residents and businesses from Congo and Côte d'Ivoire as well as KwaZulu and Zimbabwe, who thrive despite overcrowding, street crime, and uneven governance from the city (Beall, Cranshaw and Parnell 2002: 109–28).

Although a fictional series and thus a closed (because pre-recorded) narrative running over four seasons, *Gaz'lam* stages a productive tension between didactic emphasis and critique by stealth. Since its inception, the show and its spin-offs, from sound-track CDs to school-based workshops role-playing revisions of characters' behavior, have played the edge between 'aspirational' and 'sensational.' Following aspirant musician, Sifiso (newcomer Siyabonga Shibe), from rural KwaZulu to Johannesburg, it charts in generic terms the latest iteration of a familiar country-to-city story captured in films from *Jim Comes to Jo'burg* (1949) to *Jump the Gun* (1996), but modifies this rural-to-urban trajectory by having Sifiso return in the final season to his rural roots to be groomed for tribal leadership. As the publicity material puts it, the post-apartheid twist on this old story is Sifiso's 'immersion in the celebrity lifestyle' around the music scene. His education begins in gestic fashion as he is taught how to 'walk as though [he has] money,' and learns to learn from both newbies to the city, such as Khetiwe (Mbali Ntuli), his girlfriend from KwaZulu who flees an arranged marriage to the chief, and from seasoned urbanites. The latter include the singer Lerato (played by Boom Shaka's former singer Thembi Seete), whose music he mixes, and petty gangster and self-styled Ghetto Professor, GP, played by Israel Mokoe, known perhaps to overseas viewers as the father of the protagonist in *Tsotsi* (2005) (SABC *Gaz'lam Characters*, 2002). Without preaching, each episode dramatizes conflicts between the characters' aspirations and their actions, and each ends with a screen display of the local AIDS Hotline number.

The depiction of relations between people as social as well as intimate, which Brecht called 'gestic' to highlight its demonstrative, analytical, and thus instructive dimension, rather than the 'heroic realism' of the protest march, can give *Gaz'lam* a vital role in social representation.

The tension between education and entertainment plays out explicitly in the introductory documentary that accompanies DVD sales. Like the series, the introduction is the product of private sector collaboration with SABC-Education. In this video, writer-director Barry Berk, co-creator Lauren Segal, technicians, and actors in the first season speak directly to the camera in brightly lit studios to explain the educational value of dramatizing the risks of life in the edgy city, enjoyed by some and craved by other characters. These comments address the aspirations of the characters to realize their dreams of success in the big city as well as the consequences – alienation, illness, and death – of pursuing sex, money, and drugs in and around the music industry. They are juxtaposed with on-location clips of sex, drugs, and *kwaito* whose glamour is enhanced by dark filters, the music of Ghetto Ruff and other local bands, and hip clothing of the kind designed by Loxion Kulcha, which also appears in the commercial breaks.[8] The action takes place in recognizable inner-city spaces, from apartments in Yeoville to performance venues in Newtown, to the upscale offices where the two most driven female characters work; to this extent the series quotes a realist tradition of black urban life established in the anti-apartheid era and exemplified by transitional series like *The Line* (1994) (Kruger, 2006). The production design differs sharply from that era, however. Through the montage of jerky titles floating on shifting nightlight, recalling US series like *CSI (Crime Scene Investigation)*, to lushly nostalgic black and white long shots for flashbacks to rural KwaZulu, the viewer is drawn into a world of jagged edges and moral ambiguities. While the domestic scenes of confrontation and confession may echo the sentimental affect of immensely popular soap operas like the top-rated *Generations* (Kruger, 2004), the stop-action sequences of body parts, clothes, and, crucially, condoms, which characterize some sex scenes, clearly draw on the form as well as content of commercials directed at urban youth.

This appeal to the viewer as consumer does not automatically negate the educational claims of the producers, but it does highlight rather than resolve the tension. While the publicity material claims that *Gaz'lam*, 'reflects South Africa's key social issues' including 'material aspirations [; ...] sexual negotiations and betrayal; conflicts between tradition and modernity' and pays special attention to the 'HIV/AIDS pandemic that is ravaging South Africa' (SABC *Gaz'lam Series Overview*,

2002), and school groups workshop alternative scenarios, the effect of these scenarios, the 'aspirational' effect of 'buppie' (i.e., Black urban professional) success or, more pointedly, the alleged deterrent of seeing characters dying of AIDS or otherwise suffering from irresponsible behavior – what we might call in Brechtian terms, its critical dis-illusion effect – remains debatable. What makes this issue *debatable*, rather than merely dubious, is the question whether social critique can animate and not merely address its target audience *through* the commodity form of mass entertainment, whose function is usually assumed to obfuscate critique. Because critique by stealth through harnessing the transformative force of the commodity deploys cunning rather than heroic opposition, it runs the risk of commodification, or of what T. W.Adorno (1991) saw as political art's inevitable function as a deception machine for capturing, commodifying, and selling viewer time to advertisers.

But the formal and cultural nuance of the narrative, what Adorno might call its autonomy from immediately didactic purpose, allows for more complex effects. The binary oppositions that may seem pat in the series overview appear in the narrative as multi-pronged engagements that challenge the simplistic oppositions between modern and traditional that animated earlier 'naif in the city' films. Lerato's first hit single combines a 1960s *mbaqanga* track (from a popular Zulu musical *Umoja*) with 1990s hip hop sampling; Sifiso's and Khetiwe's premarital sex without penetration (*ukosoma*) suggests a traditional alternative to supposedly more sophisticated city behavior, and even the urbanized club-owner Jerome (Menzi Ngubane) asks Lerato's mother permission to live with her daughter. The most apparently urbanized character is also the most self-destructive. Although Sifiso, the naïf transformed by the city, is the evident protagonist, GP offers a clearer case of critique by stealth. The glamorous gangster has become a cliché in South African culture, from the 'American' gang who imitated Richard Widmark films in the 1950s to various screen incarnations – from *Come Back Africa* (1959) to *Mapantsula* (1988) – and GP plays to this type from the outset. He presents himself in the documentary as 'born, bred and buttered in Alex' (the slum near the rich suburb Sandton), and the direction reinforces the swagger with a close up of his elaborately layered midriff. All braggadocio in the first season, he responds to his HIV status in the second by mad swings between incipient acknowledgment and aggressive denial, which hastens his death, but GP gains depth in the third season from post-apartheid writer Sello Duiker's sense of the panic under the bravado.

The third season opened with GP already dead, but returning as a ghost. He introduces the first episode and haunts several thereafter, watching Sifiso break into and out of the music business as he wrecks several relationships and spirals into crack addiction. GP's intermittent appearances and occasional comments are ironic rather than moralizing, acknowledging audience interest in the glamour that he continues to represent, while pushing the limits of this identification with flashbacks to his AIDS-ravaged body from the second season. This play between a Brechtian sobriety and melodramatic heightening may seem unrealistic to outsiders, but the jagged extremes of this characterization offer a fictional counterpart to the actual toll of city life on members of the team: Duiker, the principal writer of the dark third season, committed suicide; Sipho Mzobe, playing Welile, Sifiso's maladjusted cousin, was murdered outside his home after concluding the first season; and Mbali Ntuli, as Khetiwe, left the series reputedly because the role, including time as a sex-worker, was too 'heavy,' even though she had tackled teenage delinquency in *Yizo, Yizo* (Motuba, 2005).

Despite GP's charismatic narration and the story's focus on Sifiso, the female characters have attracted more critical attention. When the series opened in 2002, the critical *Mail and Guardian* reiterated the producers' emphasis on the centrality of women in probing issues such as the tension between rural traditions of arranged marriage and individual aspirations (Khetiwe), between career aspirations and sexploitation in the music industry (Lerato), and the particular challenges facing single mothers (Thuli played by Bubu Mazibuko, a *Yizo, Yizo* veteran), and HIV positive women (Bonnie Mbuli as Portia). Given that audiences have been known to complain if shows portray images of particular places (such as a club or squatter camp) that differ from their known location, this kind of conversation is quite plausible. Although the acting generally follows the conventional identification of actor with character, audiences have shown a matter-of-fact response to the separation of actor and character, adapting to a new actor in an old role, such as Hlengiwe Lushaba replacing Mbali Ntuli as Khetiwe in the third season (Motuba, 2005). This practice, which would be unlikely in star vehicle soap operas, where an actor's departure would demand the death and/or disappearance of the character (as in the United States), partly reflects the lower status and possibly less reliable revenue source for the national broadcaster of a series marked educational, but partly also allows for estrangement from the actor as star while opening up space for critical reaction to the character's actions and judgments.

Seeing *through* the actor as commodity in both senses – unmasking the commodity as an illusion, while going beyond the illusion to grasp the commodity as an object of critical representation – the audience of *Gaz'lam* has so far understood the function of actors in television series generally as replaceable vehicles, and in this series, in particular, in terms of their ability to illuminate their characters' agency in the turbulent new South African city. The transformative impact of this understanding, or the degree to which the simulation and stimulation of 'people talking about sex, AIDS and life all around the country,' as Thembi Seete put it in the documentary, might direct these edgy agents from risky to responsible behavior, has yet to be fully assessed, but the critical engagement of producers and recipients in this debate suggests that in this context television, the most conventionally debased mass medium, might yet contribute to the creation of democratic agency.

Gender politics in the theatre and beyond

Theatre people have responded to the new television in part by joining it – directors such as Lara Foot-Newton and Malcolm Purkey joined the writing teams of serials *Gaz'Lam* and *Hard Copy* respectively – but this crossover belies a renewed vitality in the theatre. With Purkey as Artistic Director from 2005, the Market Theatre more than doubled door takings over the previous year under actor John Kani and ended 2005 with an accumulated surplus for the first time since 1986 (Market Theatre Foundation, 2005: 1). Audiences responded not only to the renovation of the Newtown Precinct that had added retail and residential developments as well as more effective security to its cultural spaces, but also to a new repertoire that moved beyond nostalgia to plays that tackled actual South African conflicts and controversies. Where other venues, such as the Civic Theatre, mimic the casinos with spectacles like *Umoja: Spirit of Togetherness* (2005), the latest in a long line of Zulu spectacles, the Market Theatre grapples with contemporary South Africa. In 2005 it staged plays that locate their drama in current South African controversies from the rural, such as the risks of botched initiation rites in Mncedise Thambe's *Esuthwini*, to the urban, such as the violence pervading the inner-city sex and drugs trade in Mpumelelo Paul Grootboom's *Cards*. These plays may be uneven, but they raise compelling themes and sketch out forms that might, in more practiced hands, prove equal to the complexities of the present moment.[9]

The 'political thriller' by satirist Mike van Graan offers a good indication. His *Green Man Flashing* (2004) deals with characters caught between anti- and post-apartheid pressures. The plot of corruption and rape by a Zulu politician uncannily anticipated the 2006 rape trial of former deputy president and '100% Zuluboy' Jacob Zuma (Moya, 2006). The debate continued as feminists analyzed the impact of Zuma's behavior and the 'kangaroo court' on South African sexual politics (Motsei, 2007a, b); this debate has only become more urgent since the Zuma faction ousted President Mbeki before the latter's term ended in 2009 and Zuma himself became president in April of that year.[10] Although it deals ostensibly with the political maneuvering that dominated headlines in the months leading up to the 1999 elections and Thabo Mbeki's assumption of the presidency after Nelson Mandela, it highlights ongoing tensions between accountability and corruption, of party loyalty and the rule of law, of male sociability in power and persistent violence against women, which challenges the present Mbeki administration as well. The play dramatizes the case of one Gabby Anderson, a 40-plus white woman who spent the late apartheid years of the 1980s in exile so that she and her black husband Aaron Matshoba might join the African National Congress (ANC) in the anti-apartheid struggle, and, more personally, live together in a manner denied by the 'Immorality' Act. Once returned, her marriage to Aaron unravels under pressure from the death of their son, stabbed by street toughs who wanted his bicycle, and Aaron's absence from home as the new government's chief mediator in conflict zones.

The central conflict erupts when Gabby is raped by her boss, Shadrack Khumalo, a powerful kingpin of the ANC faction controlling the fractious province of KwaZulu-Natal, the province torn by political violence in the run-up to the 1994 election and still troubled by instability. Aaron is sent by the party to persuade his now ex-wife to keep quiet in the name of 'national unity;' he offers her an overseas consular appointment in return for dropping all charges but also intimates, through the analogy of the 'green man flashing,' that, should she still choose to exercise her right to her day in court and thus enter the political danger zone, she might be hit by illegal but nonetheless unstoppable vehicles running red lights. The play provides an uncommonly candid view of the fall of anti-apartheid ideals of justice for all into the pit of political squabbles in which the narrow interests of the party elite are dressed up as urgent national demands. It provoked fierce debate among audiences on topics ranging from the abuse of power to the phenomenally high rape rate. Although it follows a complicated political plot that might confuse non-South Africans, *Green Man Flashing* also offers a stark picture of post-apartheid realities by blasting away the haze of anti-apartheid nostalgia in Lee Hirsch's *Amandla – a Revolution in Four Part Harmony* (2002) or

Pascale Lamche's *Sophiatown* (2004), or the sentimentality of local Oscar bait like Gavin Hood's *Tsotsi* (2005).

The Johannesburg staging of *Green Man Flashing* (directed by Clare Stopford, 2005) uses flashbacks and the differentiation of downstage 'public' space – including not only the inquest but also the intrusion into Gabby's home by Aaron and Luthando, and, with them, party politics – and upstage intimate space to which Gabby retreats, sometimes alone, sometimes in the company of Anna, shifting the action from the inquest into the death of Luthando, apparently shot by Gabby in self-defense, at which Aaron claims to have found Luthando dead after hearing three shots from outside the house, to a scene in which Luthando attempts to bully Gabby. Before the truth comes out near the end of the play, successive flashbacks take Gabby's and Aaron's quarrels and the death of their son, investigated by Abrahams, alongside Abrahams's confession to the TRC, in which he obliquely implicates Luthando as one of the black policemen implicated in torture. By the time Luthando replays his threat to Gabby, the play has also revealed that the ANC accepted Khumalo's claim that he had only consensual sex with Gabby and that Aaron is torn between Gabby's rights and his party loyalty. When Gabby shoots Luthando in the shoulder, Aaron rushes in, kills Luthando with two more shots, and argues her into the self-defense argument on the grounds that 'the world is a better place without "comrade" Luthando' (Van Graan, 2004: 54). After the shooting, Abrahams returns, having previously surprised Aaron and Luthando in Gabby's absence when he came to collect her rape statement, and offers Gabby the tape he had secretly planted earlier; he leaves her with the option of destroying the evidence and the laconic remark 'it [secret police bugs and perjured statements] feels like old times' (ibid.: 56). Despite initially agreeing to lay charges, Gabby takes up the ANC offer, arguing that 'I'm tired of feeling guilty all the time' (ibid.: 58), thus scotching the political potential of a 'high profile case' that in Anna's view might 'strike a blow for all those women, teenagers, girls, [...] babies that are raped everyday' (ibid.: 58) and, by implication, threatened with HIV. But *Green Man Flashing* ends not with this apparent resignation, but with the inquest and characters' and audience's attention once again on Gabby as she prepares to give evidence as if by video link. Although Douglas looked straight ahead as though at a camera, the stage directions suggest a more complex conclusion: acting *'as if she is physically at the inquest, [as she] takes in Abrahams, and finally Aaron, then looks away from Aaron,'* before asking rhetorically 'Where shall I start?' (ibid.:62), Gabby may follow the green light after all. The look to the imagined camera not only alludes to a fictional courtroom but also invites comparison between theatrical and televisual mediation of political agency.

In 2005, during the second and, according to the Constitution, final term of Mbeki's presidency, *Green Man Flashing* turned out to be even more topical than expected. The audience ranged widely in age and color, and everyone from the private school students (bussed in for the occasion) to those closer in age to the protagonists appeared to be vociferously debating the controversial points of the plot and likely scenarios for the ending in the lobby bar after the performance. Reviewers, who included newcomers as well as veteran critics, noted not only the topical import of its plot but also the reaction it provoked, especially in younger audiences, who may have been too young to experience the anti-apartheid struggle, but were clearly able to appreciate the play's critique of the politics as usual bargaining that has replaced the ideals of the struggle.[11] In 2006 and 2007, well after the play had closed, its controversial themes had a compelling echo in the Zuma trial and its aftermath in ongoing acrimonious debate about sexual exploitation. Beyond the immediate act of life imitating art, it created analytic space on the sticky terrain between aspiration and sensation. The example of *Green Man Flashing* may not awaken an immediate response, but its provocation of debate on uncomfortable questions about the government's toleration of sexual exploitation set a precedent that still compels attention.

From the perspective of mid-2009, as the country faces in Zuma a president clouded by scandal but as yet unchallenged by any serious opposition, and in the SABC a national broadcaster increasingly noted for executives' conspicuous consumption rather than the critical content production, the urgent social questions raised by stage plays like *Green Man Flashing* or *Cards* and by television serials like *Gaz'Lam* still demand attention long after the play and the serial have ended their runs. The struggle of women and sexual minorities to create and inhabit space that provides them with the democratic agency promised by South Africa's unique constitutional provisions for freedom of sexual orientation as well as gender equality may not dominate the media, but its representation in new serials such as *After Nine* shows creators and audiences responding critically to education by stealth. While the news department of the SABC has become even more of a government mouthpiece since the Freedom of Expression Institute published its critique (Duncan, 2001), fictional series like *After Nine* join plays at institutions like the Market in showing the potential for the critical performance of public agency in unlikely contexts, which a democratic society needs to prosper democratically.

Notes

1. For diverse views of the decline and revival of Johannesburg from the 1970s to the present, see also Tomlinson et al. (2003). For the impact of these changes on theatre and film in the city, see Kruger (2001, 2006).

2. Although critics have challenged the city government's neoliberal rhetoric of entrepreneurship, public/private partnerships – in place since 2000 – have borne fruit. In particular, the non-profit facilitator, the Central Johannesburg Partnership (CJP; since 1992) has created City Improvement Districts such as Newtown, which now includes affordable housing and retail business alongside the Cultural Precinct.

3. After struggling with too many demands, the Department of Arts, Culture, Science and Technology (DACST) split in 2002 to create a separate department of arts and culture (DAC). Although drawn from public nominations, NAC board members are appointed by the minister and have tended to channel funding to their own institutions. For the history of mismanagement, see Kruger (2002); for ongoing analysis, see the 'artwit' column by Mike van Graan: http:www.artslink.co.za.

4. As in the NAC, NFVF officers are nominated by peers but appointed by the minister. Unlike the NAC, the NFVF is managed as a not-for-profit corporation, but this structure has not prevented similar problems with mismanagement (see Dawes and Reddy, 2005). For local-content quotas, see Laschinger (2005).

5. Purkey made his name as director of the anti-apartheid Junction Avenue Theatre Company and co-creator of its Brechtian history plays, such as *Randlords and Rotgut* (1978) and *Sophiatown* (1986: revived in 1994 and 2005); McCarthy has acted in several Market plays and critical television shows, and written scripts like the second season of *Gaz'lam* (2003). In addition to *Gaz'lam* episodes, Duiker wrote the novels *13 Cent* (2000), *The Quiet Violence of Dreams* (2001), and *The Hidden Star* (2006).

6. Sexually active young women may be among the most vulnerable, but young men have also died, including sons of former president Nelson Mandela. NAPWA (National Association of People with AIDS) and the Treatment Action Campaign (TAC) remain skeptical of didactic attempts at behavior modification, preferring to lobby for increased access to treatment and better living and working conditions. For comments on the limits of behavior modification programs by the director of DramAide, see Dalrymple (2005: 163–6); for former president Mbeki's links to AIDS denialists, see Zackie Achmat (2004).

7. Mayor Aaron Masondo's bold comparison with first world global cities has provoked skepticism from local researchers such as Bremner (2004: 78–9) or expatriates such as Jennifer Robinson, who argues that the 'regulating fiction of the global city' privileges a small group of first world financial markets; she suggests instead that Johannesburg's combination of high-tech and a formal stock exchange on the one hand, and low-tech and informal but no less transnational trade on the other offers a model for rethinking urban form and transformation beyond the dichotomy between 'global success' and 'developmentalism,' in the form of ordinarily worldly rather than exceptionally global cities (Robinson, 2003: 273).

LIVERPOOL JOHN MOORES UNIVERSITY
LEARNING SERVICES

8. In the second season (2003), the roles of writer and director were split with Alex Yazbek directing and head-writer Neil McCarthy working with writer Nonhlanhla Dlamini as well as stage-director and adaptor Lara Foot-Newton. For the third (2004–05), Sello Duiker was primary writer until his death in late 2004. Loxion Kulcha's founder designers are old enough to remember apartheid, but young enough to speak to the 2000 generation. The language and style of these labels borrow from hip hop in their contraction of words, here 'location [former township] culture.'

9. Grootboom's next play, *Relativity* (at the State Theatre in Pretoria in 2006) delivered more of the same, portraying not only a serial killer rapist strangling victims with their G-strings, but also relentless domestic violence. Although he denies charges of glamorizing violence and especially of titillating male spectators, his admiration for Quentin Tarantino casts doubt on his claim to depict the 'violent and unforgiving society' as it 'actually happens' in his own neighborhood (quoted in Botha, 2005).

10. Zuma was initially dismissed from the deputy presidency for accepting payments from a corrupt associate who was convicted in 2005, but Zuma's trial for raping a family friend half his age galvanized support from the ANC Youth League, and some elements in the Congress of South African Trade Unions (COSATU) dissatisfied with the neoliberal line of Mbeki's government. Zuma may have been an unlikely leader of grassroots opposition since, like Mbeki, he spent the apartheid years in exile without the bracing democracy of the anti-apartheid movement. Nonetheless, his counsel was able to get the Friends of Jacob Zuma access to the court and media attention. Its Zulu name, *Awulethu uMshini wam* (lit. 'bring me my machine'), in its current extension, 'machine gun,' is most likely a celebration of his alleged sexual prowess. In contrast, the judge, a white man who took the case after three Zulu judges recused themselves, denied access to anti-rape organizations such as the long-standing People Opposing Women Abuse (POWA). Zuma was acquitted of the rape charge in May 2006 but berated by the judge for having unprotected sex and for making the claim, astonishing for the former chair of the National AIDS Council, that the HIV virus could be eliminated by taking a shower. He was to have been tried for corruption in July 2006, but the high court judge lost patience with the prosecution's delays and ordered the case thrown out of court until the National Prosecution Authority was ready to submit a full indictment. In April 2009, shortly before the general election, the NPA finally dropped charges against Zuma on the grounds that the prosecution had compromised the case with questionable legal tactics (see *Mail and Guardian*, 2009).

11. See, for instance, Rafiek Mammon in the *Cape Town Argus* (8 July 2004, cited in Van Graan, 2004: 64): 'The first people who [... gave the play] its deserved standing ovation [...] were a group of [...] young adults.'

References

Achmat, Z. 'The Treatment Action Campaign, HIV/AIDS and the government,' *Transformation: Critical Perspectives on Southern Africa*, 54:1 (2004) 76–84.

Adorno, T. W. *The Culture Industry: Selected Essays on Mass Culture* (London and New York: Routledge, 1991).

Auslander, P. *Liveness: Performance in a Mediatized Culture* (London and New York: Routledge, 1999).

Banham, M. et al. (eds) *A History of African Theatre* (Cambridge: Cambridge University Press, 2004).

Baumol, W. J. and W. G. Bowen. *Performing Arts: The Economic Dilemma: A Study of Problems Common to Theatre, Opera, Music and Dance* (Cambridge, MA: Massachusetts Institute of Technology Press, 1966).

Beall, J., O. Cranshaw, and S. Parnell. *Uniting a Divided City: Governance and Social Exclusion in Johannesburg* (London: Earthscan, 2002).

Beavon, K. *Johannesburg: The Making and Shaping of the City* (Pretoria: University of South Africa Press, 2004).

Botha, N. 'Mad, bad and relative,' *Johannesburg Mail and Guardian* (17 October 2005).

Bourgault, L. *Playing for Life: Performance in Africa in the Age of AIDS* (Durham, NC: Carolina Academic Press, 2003).

Brecht, B. *Werke: Große kommentierte Berliner und Frankfurter Ausgabe* (Frankfurt: Suhrkamp, 1998).

———. *Brecht on Film and Radio,* trans. M. Silberman (London: Methuen, 2000).

———. *Brecht on Art and Politics,* trans. T. Kuhn and S. Giles (London: Methuen, 2002).

Bremner, L. *Johannesburg: One City, Colliding Worlds* (Johannesburg: STE Publishers, 2004).

City of Johannesburg 'Vision' (Chapter Five of *iGoli 2030*), 2002a: http://www.joburg.org.za/feb_2002/2030-vision.pdf (accessed 7 June 2005).

———. *iGoli 2030: Short Version,* 2002b: http://www.joburg.org.za/feb_2002/2030-shortversion.pdf (accessed 7 June 2005).

DACST (Dept of Arts, Culture, Science and Technology 1994–2002) *Creative South Africa: A Strategy for Realising the Potential of the Cultural Industries* (Pretoria; Government: Publications, 1998).

Dalrymple, L. 'Drama Studies in the 21st Century,' *South African Theatre Journal,* 19 (2005) 157–72.

Dawes, N. and S. Reddy. 'Lights, Camera, Showdown,' *Mail and Guardian* (4 November 2005).

Duncan, J. *Broadcasting and the National Question: South African Broadcast Media in an Age of Neo-Liberalism* (Johannesburg: Freedom of Expression Institute, 2001).

Fraser, N. 'Inner City Regeneration Overview for 2005,' *Joburg* (19 December 2005).

Green, P. 'The Rise and Fall of the SABC,' *Mail and Guardian* (27 July 2007).

HSRC (Human Sciences Research Council, South Africa) *National HIV Prevalence, Incidence, Behavior and Communication,* 2005: http://www/hsrc.ac.za/media/2005/11/20051130_1Factsheet2.html (accessed 4 April 2006).

Kruger, L. *The Drama of South Africa: Plays, Pageants and Publics since 1910* (London and New York: Routledge, 1999).

———. 'Theatre, Crime and the Edgy City in Post-Apartheid Johannesburg,' *Theatre Journal,* 53:2 (2001) 223–52.

———. 'Scarcity, Conspicuous Consumption and Performance in South Africa,' *Theatre Research International*, 27:3 (2002) 232–42.

———. 'Theatre for Development and TV Nation: Educational Soap Opera in South Africa,' *African Drama and Performance*, ed. J. Conteh-Morgan and T. Olaniyan (Indianapolis: Indiana University Press, 2004), pp. 155–75.

———. 'Filming the Edgy City: Cinematic Narrative and Urban Form in Post-Apartheid Johannesburg,' *Research in African Literatures*, 37:2 (2006) 141–63.

———. 'Letter from Johannesburg: Performance and Urban Fabrics in the Inner City,' *Theater*, 38:1 (2008) 1–17.

Laschinger, K. 'Indigenous Rewards,' *The Media Online* (December 1, 2005).

Mail and Guardian, 'Mpshe: Zuma decision not an acquittal,' *Mail and Guardian* (6 April 2009).

Marlin-Curiel, S. 'Wielding the Cultural Weapon After Apartheid; Bongani Linda's Victory Sonqoba Theatre in South Africa,' in *Theatre and Empowerment: Community Theatre on the World Stage*, ed. R. Boon and J. Plastow (Cambridge: Cambridge University Press, 2004), pp. 94–123.

Market Theatre Foundation, *2004–2005 Annual Report* (Johannesburg, 2005): http://www.markettheatre.co.za (accessed 24 March 2006).

Mofokeng, J. 'Theatre for Export: The Commercialization of the Black People's Struggle in South African Export Musicals,' in *Theatre and Change in South Africa*, ed. G. Davis and A. Fuchs (Amsterdam: Overseas Publishers Association, 1996).

Motsei, M. *The Kanga and the Kangaroo Court: Reflections on the Rape Trial of Jacob Zuma* (Johannesburg: Jacana, 2007).

———. 'Jacob Zuma's bankrupt sexual morality afflicts millions of us,' *Johannesburg Sunday Times*, News and Opinions section (29 April 2007) 21.

Motuba, I. 'A Tough Act to Follow,' *Johannesburg Star (Tonight)* (12 April 2005).

Moya, F-N. '100% Zuluboy,' *Mail and Guardian* (6 April 2006).

NAC (National Arts Council). *National Arts Council of South Africa: General Information*, 2005: http://www.nac.org.za/nac.htm (accessed 1 November 2005).

NFVF (National Film and Video Foundation). *NFVF Overview*, 2005: http://www.nfvf.co.za/ overview.htm (accessed 1 November 2005).

Phelan, P. *Unmarked: The Politics of Performance* (London and New York: Routledge, 1993).

Purkey, M. 'Market Forces: Interview with Loren Kruger,' *Theater*, 38:1 (2008) 18–29.

Robinson, J. 'Johannesburg's Futures: Beyond Developmentalism and Global Success,' in *Emerging Johannesburg*, ed. R. Tomlinson et al. (London and New York: Routledge, 2003), pp. 259–80.

Schechner, R. *Performance Theory* (London and New York: Routledge, 1988).

SABC (South African Broadcasting Corporation). *Gaz'lam Series Overview* (season 2), 2002: http://www/sabceducation.co.za/gazlam/Gaz_series.html (accessed 30 June 2005; no longer available).

———. *Gaz'lam: Characters* (season 1 and 2), 2002: http://www.sabceducation.co.za/gazlam/Gazlam_Characters.html (accessed 30 June 2005; no longer available).

Simone, A. 'The Visible and Invisible: Remaking Cities in Africa,' in *Under Siege: Four African Cities – Dokumenta 11: Platform 4*, ed. O. Enwezor et al. (Kassel: Hatje Cantz, 2002), pp. 23–44.

Tomlinson, R. et al. (eds) *Emerging Johannesburg: Perspectives on the Post-Apartheid City*. (London and New York: Routledge, 2003).

Van Graan, M. *Green Man Flashing* (Cape Town: n.p., 2004).

———. 'Towards a Free Market' (interview with Sibongiseni Mkhize), *Mail and Guardian* (14 January 2005).

Index